LECTURES ON GENERAL RELATIVITY

LECTURES ON GENERAL RELATIVITY

by

A. PAPAPETROU

Laboratoire de Physique Théorique, Institut Henri Poincaré,
11 Rue Pierre et Marie Curie, Paris 5, France

D. REIDEL PUBLISHING COMPANY

DORDRECHT-HOLLAND / BOSTON-U.S.A.

Library of Congress Catalog Card Number 74–81943

Cloth edition: ISBN 90 277 0514 3
Paperback edition: ISBN 90 277 0540 2

Published by D. Reidel Publishing Company,
P.O. Box 17, Dordrecht, Holland

Sold and distributed in the U.S.A., Canada, and Mexico
by D. Reidel Publishing Company, Inc.
306 Dartmouth Street, Boston,
Mass. 02116, U.S.A.

Printed in The Netherlands by D. Reidel, Dordrecht

TABLE OF CONTENTS

CHAPTER IV / THE GRAVITATIONAL FIELD

CHAPTER V / THE SCHWARZSCHILD SOLUTION

CHAPTER VI / SOME OTHER EXACT SOLUTIONS

CHAPTER VII / WEAK GRAVITATIONAL FIELDS

CHAPTER VIII / VARIATIONAL PRINCIPLE. IDENTITIES, CONSERVATION LAWS

PREFACE

This book is an elaboration of lecture notes for the graduate course on General Relativity given by the author at Boston University in the spring semester of 1972. It is an introduction to the subject only, as the time available for the course was limited.

The author of an introduction to General Relativity is faced from the beginning with the difficult task of choosing which material to include. A general criterion assisting in this choice is provided by the didactic character of the book: Those chapters have to be included in priority, which will be most useful to the reader in enabling him to understand the methods used in General Relativity, the results obtained so far and possibly the problems still to be solved.

This criterion is not sufficient to ensure a unique choice. General Relativity has developed to such a degree, that it is impossible to include in an introductory textbook of a reasonable length even a very condensed treatment of all important problems which have been discussed until now and the author is obliged to decide, in a more or less subjective manner, which of the more recent developments to omit. The following lines indicate by means of some examples the kind of choice made in this book.

There is a relatively long chapter on gravitational radiation, describing the different aspects of the problem which have been discussed until now. The problem of motion is discussed in a short chapter, just sufficient to show that in General Relativity the equations of motion are determined by the field equations. An equally short chapter on cosmology deals with the general consequences drawn from the cosmological principle combined with the field equations of General Relativity. Important problems such as the Cauchy problem as well as the quantization of the gravitational field had to be omitted entirely.

No attempt has been made to give references to original papers; only a small number of such references could be included and deciding which references to give and which to omit is a nearly impossible task.

Thanks are due to Professor John Stachel for several discussions concerning the organization of the lectures. I am also grateful to Professor Werner Israel for some very useful suggestions.

Special thanks are due to Dr John Madore who read the entire manuscript. He made a number of suggestions leading to essential improvements and helped eliminate several mistakes as well as the more important linguistic shortcomings of the text.

Paris, September 1974 A. PAPAPETROU

INTRODUCTION

A student knowing the special theory of relativity will understand at once that general relativity is some generalization of the special theory. What is not indicated directly is that general relativity is a relativistic theory of the gravitational field. In fact, general relativity has several aspects, related to different problems of physics as well as of geometry. We shall first give a qualitative description of the content of this theory. This will be followed by a short survey of the historical development of the theory of the gravitational field and of the present situation in general relativity.

The characteristic feature of special relativity is the rejection of the concepts of absolute space and time, used in prerelativistic physics. This has as an immediate consequence the impossibility of defining the absolute velocity of a body. Meaningful is only the relative velocity of one body with respect to another. However, in special relativity we still have an infinite set of inertial coordinate systems, all being equivalent to each other. The basic property of the inertial coordinate systems is the same as in Newtonian mechanics: In such a coordinate system a particle moving without any forces acting on it has a constant velocity, or in other words it has zero acceleration. A particle having zero acceleration in one inertial coordinate system has the same property in all inertial coordinate systems. Similarly a particle having non-zero acceleration in one inertial coordinate system has the same property in all inertial coordinate systems. This means that in the special theory of relativity the acceleration continues to have an absolute meaning.

In general relativity the acceleration also loses its absolute meaning. This is equivalent to the statement that in general relativity there are no inertial or otherwise preferred coordinate systems: *All coordinate systems are equivalent.* This is actually the reason for which the theory is called general relativity.

From the mathematical point of view the equivalence of all coordinate systems demanded by general relativity is obtained in a straightforward manner. The Minkowski space of special relativity is the simplest form of a *Riemannian space*: it is a *flat* space and therefore it allows a set of geometrically preferred coordinate systems, which are the inertial frames. In general relativity the Minkowski space is replaced by a *curved* Riemannian space, in which there are generally no preferred coordinate systems.

From the physical point of view the transition from flat to a curved Riemannian space has been found necessary in order to obtain a quantitative description of the gravitational field. Indeed, in general relativity the gravitational field is only the expression of the fact that we have in reality a curved Riemannian space and not the flat Minkowski space. We may therefore say that the characteristic feature of general relativity is the geometric description or simply the *geometrization* of the gravitational field.

Historically the first theory of the gravitational interaction is the one formulated by Newton. Of course this theory uses the concepts of absolute space and time, which form the basis of Newtonian mechanics. The Newtonian theory of gravitation has been extremely successful in the discussion of the planetary motions: The order of magnitude of the discrepancies between the theoretical predictions and the observations is characterised by an unexplained residue in the motion of the perihelion of Mercury, which has the extremely small value of about 40″ per century.

In the beginning of the 20th century the system of physical concepts and ideas was modified profoundly by the special theory of relativity and this theory had to be accepted, because of its impressive successes, as a basic part of physics. Since the concepts of absolute space and time have been definitively rejected by special relativity, it became clear that the Newtonian theory of gravitation had to be replaced – in spite of all its successes – by a new theory using the Minkowski space-time.

Several attempts were made, and are still being made, to formulate a relativistic theory of the gravitational field within the framework of special relativity. We shall mention here only the theory of Nordström (1912), as this was the first logically consistent theory of this type. The theory of Nordström was found to be successful in the following sense: In a very good first approximation it reduces to the Newtonian theory of gravitation and thus it incorporates all the successes of the Newtonian theory. However, in the second approximation it gave a correction to the rotation of the perihelion of Mercury which was wrong even in the sign.

After long systematic efforts Einstein arrived at the conclusion that for a satisfactory description of gravitation one had to modify also the special theory of relativity and to replace the Minkowski space by a curved Riemannian space. This led to general relativity, which is the subject of these lectures. We shall mention already here the fact that, contrary to the Newtonian theory as well as to the theory of Nordström, which are *scalar* theories, general relativity is a *tensor* theory of the gravitational field: The field is described by the metrical tensor of the Riemannian space, its source being the matter tensor. Note that all components of the matter tensor are acting effectively as sources of the gravitational field in a detailed manner determined by the field equations.

This amazing theory, resulting from the postulate of interdependence of the concepts of space on the one side and matter on the other, previously considered as totally independent, could claim a number of successes immediately after its formulation. Indeed, general relativity reduces in the first or *linear* approximation to the Newtonian theory of gravitation, thus incorporating the successes of this theory. In the second approximation it explains quantitatively the residue in the rotation of the perihelion of Mercury. It also predicts several other small effects, two of them discussed in the very beginning by Einstein: The deflection of light in a gravitational field and the gravitational red shift of the spectral lines emitted by atoms in the vicinity of large masses. These effects have been verified subsequently by the observations, at least qualitatively.

Until about 1950 the problems discussed in general relativity all lead to extremely small relativistic effects, always near or under the limit of present possibilities of observation. Thus general relativity seemed to be rather unimportant from the practical point of view. The only exception has been the cosmological problem, for the discussion of which the equations of general relativity gave entirely new possibilities leading to important and unexpected new results.

A new situation has been created by the discovery of the possibility of gravitational collapse, predicted by the equations of general relativity for sufficiently large masses satisfying certain conditions. This discovery stimulated the systematic development of relativistic astrophysics, which is certainly the most discussed, at present, application of general relativity.

We shall mention at this point two other important problems in general relativity. Gravitational radiation has been discussed theoretically from several aspects, but a sufficiently coherent picture of the situation is still lacking. The experimental verification of the existence of gravitational radiation will evidently be crucial for a final proof of the correctness of general relativity; however, it seems that decisive results will not be obtained in the near future. As a last problem we mention the quantization of the gravitational field – an important but also an extremely difficult problem.

We shall close this introduction with two technical remarks. It will be assumed in these lectures that the reader is acquainted with the concepts and the methods of the special theory of relativity. On the other hand, tensor calculus in general coordinates and Riemannian geometry will be developed here to the extent needed.

CHAPTER I

TENSOR CALCULUS

1. Scalars and Vectors. Tensors

Any physical quantity, e.g. the velocity of a particle, is determined by a set of numerical values – its components – which depend on the coordinate system. Studying the way in which these values change with the coordinate system leads to the concept of *tensor*, which we shall develop in this chapter. With the help of this concept we can express the physical laws by *tensor equations*, which have the same form in any coordinate system.

In an intuitive approach we may say that the two basic concepts of tensor calculus are the *scalar* and the (contravariant) *vector*.

A scalar is any physical quantity determined by a single numerical value – i.e. having just one component – which is independent of the coordinate system. Examples of scalar: In Newtonian mechanics the mass or the charge of a particle; in special relativity the rest-mass and again the charge of a particle.

The simplest example of a vector is a displacement **AB** in e.g. ordinary Euclidean space. If we use Cartesian coordinates and if the coordinates of the points A and B are $x_A^{\dot{\mu}}$ and $x_B^{\dot{\mu}}$ ($\dot{\mu}=1, 2, 3$), then the components of the vector **AB** are $x_B^{\dot{\mu}}-x_A^{\dot{\mu}}$. This is, however, a very special case. When we use non-Cartesian coordinates we can define in this way only an infinitesimal displacement, leading from the point A with coordinates x^μ to a neighbouring point A' with coordinates $x^\mu+\mathrm{d}x^\mu$. The components of such a vector are the differences of the coordinates of the points A and A', i.e. the differentials $\mathrm{d}x^\mu$.

Note that the definition of the infinitesimal displacement vector **AA**$'$ is now valid for general coordinates and also in any (flat or non-flat) space having an arbitrary dimension n. In general relativity we shall be interested in the case $n=4$. However, until we come to the applications of tensor calculus to general relativity we shall leave the n unspecified.

Starting from the infinitesimal vector **AA**$'$ with components $\mathrm{d}x^\mu$ we can construct a finite vector v^μ defined at the point A in the following way. Consider a smooth curve passing through the points A and A'. This curve will be determined by n functions of a scalar parameter λ:

$$x^\mu = f^\mu(\lambda).$$

If the points A and A' correspond to the values λ and $\lambda+\mathrm{d}\lambda$ of the parameter, then the *tangent vector* v^μ to the curve at A has the components

$$v^\mu = \frac{\mathrm{d}x^\mu}{\mathrm{d}\lambda}. \tag{1.1}$$

The infinitesimal displacement $\mathrm{d}x^\mu$ or equivalently the vector v^μ will be used as the prototype of a *contravariant vector*. Let us now ask the following question: If we consider a coordinate transformation leading from the initial coordinates x^μ to new coordinates $\tilde{x}^\mu$, what will be the components of the vector **AA**′ in the new coordinate system? The answer to this question will be found at once if we recall that any transformation $x^\mu \to \tilde{x}^\mu$ is determined by n equations of the form:

$$\tilde{x}^\mu = f^\mu(x^\alpha); \qquad \mu, \alpha = 1, 2, \ldots, n. \tag{1.2}$$

From these equations we derive at once the relation

$$\mathrm{d}\tilde{x}^\mu = \sum_\alpha \frac{\partial \tilde{x}^\mu}{\partial x^\alpha} \mathrm{d}x^\alpha. \tag{1.3}$$

Equation (1.3) gives the components $\mathrm{d}\tilde{x}^\mu$ in the coordinate system $\tilde{x}^\mu$ of the infinitesimal displacement **AA**′ which has the components $\mathrm{d}x^\alpha$ in x^α.

We can derive from (1.3) the transformation law for the vector v^μ given by (1.1), if we take into account that the parameter λ is a scalar: The differential $\mathrm{d}\lambda$ has the same value in both coordinate systems and consequently we find at once:

$$\tilde{v}^\mu = \sum_\alpha \frac{\partial \tilde{x}^\mu}{\partial x^\alpha} v^\alpha. \tag{1.3′}$$

We can now give the general definition of a contravariant vector a^μ: It is a quantity with n components depending on the coordinate system in such a way that the components a^μ in the coordinate system x^μ are related to the components $\tilde{a}^\mu$ in $\tilde{x}^\mu$ by a relation of the form (1.3) or (1.3′):

$$\tilde{a}^\mu = \sum_\alpha \frac{\partial \tilde{x}^\mu}{\partial x^\alpha} a^\alpha. \tag{1.4}$$

An important remark has to be made about the transformation formula (1.2), or equivalently about Equation (1.3). In order to have coordinate systems x^μ and $\tilde{x}^\mu$, which are equally acceptable, we must be able to solve Equation (1.3) with respect to $\mathrm{d}x^\alpha$. This will be possible only if

$$\det \frac{\partial \tilde{x}^\mu}{\partial x^\alpha} \neq 0 \quad \text{or} \quad \infty. \tag{1.5}$$

When the condition (1.5) is satisfied at the point A, we say that the transformation is regular at A. Remembering now the relations

$$\sum_{\alpha} \frac{\partial \tilde{x}^{\mu}}{\partial x^{\alpha}} \frac{\partial x^{\alpha}}{\partial \tilde{x}^{\lambda}} = \delta^{\mu}_{\lambda} = \sum_{\alpha} \frac{\partial x^{\mu}}{\partial \tilde{x}^{\alpha}} \frac{\partial \tilde{x}^{\alpha}}{\partial x^{\lambda}}; \tag{1.6}$$

$$\delta^{\mu}_{\lambda} = 1 \quad \text{if} \quad \mu = \lambda, \qquad \delta^{\mu}_{\lambda} = 0 \quad \text{if} \quad \mu \neq \lambda, \tag{1.7}$$

we find at once that the solution of (1.3) for $\mathrm{d}x^{\alpha}$ is:

$$\mathrm{d}x^{\alpha} = \sum_{\mu} \frac{\partial x^{\alpha}}{\partial \tilde{x}^{\mu}} \mathrm{d}\tilde{x}^{\mu}. \tag{1.8}$$

The *Kronecker symbol* δ^{μ}_{λ} plays an important role in tensor calculus as we shall see later.

Using the concepts of scalar and of contravariant vector we can define a *covariant vector* b_{μ}, written with a lower index μ, in the following way: The n components b_{μ} shall depend on the coordinate system in such a way that, if a^{μ} is *any* contravariant vector, the sum $\sum_{\mu} b_{\mu} a^{\mu}$ is a scalar:

$$\sum_{\mu} b_{\mu} a^{\mu} = \sum_{\mu} \tilde{b}_{\mu} \tilde{a}^{\mu} \quad \text{for any} \quad x^{\mu} \to \tilde{x}^{\mu}. \tag{1.9}$$

The transformation formula for b_{μ} can be derived at once from (1.9) and (1.4). We find:

$$\sum_{\alpha} a^{\alpha} \left(b_{\alpha} - \sum_{\mu} \frac{\partial \tilde{x}^{\mu}}{\partial x^{\alpha}} \tilde{b}_{\mu} \right) = 0.$$

Since this relation must hold for any vector a^{α}, i.e. for any set of numbers a^{α}, we must have:

$$b_{\alpha} = \sum_{\mu} \frac{\partial \tilde{x}^{\mu}}{\partial x^{\alpha}} \tilde{b}_{\mu}. \tag{1.10}$$

Equation (1.10) is the transformation formula for a covariant vector. Using (1.6) we can solve (1.10) for $\tilde{b}_{\mu}$. The result is:

$$\tilde{b}_{\mu} = \sum_{\alpha} \frac{\partial x^{\alpha}}{\partial \tilde{x}^{\mu}} b_{\alpha}. \tag{1.10'}$$

The sum (1.9) is called the *scalar product* of the vectors a^{μ} and b_{μ}.

To simplify the formulae we shall use henceforth the *summation convention* introduced by Einstein:

$$T^{\cdots}_{\cdots\alpha} V^{\cdots\alpha}_{\cdots} \equiv \sum_{\alpha} T^{\cdots}_{\cdots\alpha} V^{\cdots\alpha}_{\cdots}. \tag{1.11}$$

I.e. we shall sum over an index α, if in any expression the letter α appears twice, once as an upper and once as a lower index. Thus we shall write instead of (1.10′):

$$\tilde{b}_\mu = \frac{\partial x^\alpha}{\partial \tilde{x}^\mu} b_\alpha .$$

With the additional remark, suggested by the relation (1.10′), that an upper index in the denominator is equivalent to a lower index we shall write also instead of (1.4):

$$\tilde{a}^\mu = \frac{\partial \tilde{x}^\mu}{\partial x^\alpha} a^\alpha .$$

The definition of tensors of higher order is straightforward. A contravariant tensor of order two is a quantity having n^2 components $T^{\lambda\mu}$ – the indices λ and μ taking independently the values 1, 2, ..., n – which transforms, when $x^\mu \to \tilde{x}^\mu$, in such a way that, if a_μ and b_μ are arbitrary covariant vectors, the sum $T^{\lambda\mu} a_\lambda b_\mu$ is a scalar:

$$T^{\lambda\mu} a_\lambda b_\mu = \tilde{T}^{\lambda\mu} \tilde{a}_\lambda \tilde{b}_\mu \quad \text{for any transformation } x^\mu \to \tilde{x}^\mu . \tag{1.12}$$

Similarly we define a covariant tensor $T_{\lambda\mu}$ of order two by the condition:

$$T_{\lambda\mu} a^\lambda b^\mu = \tilde{T}_{\lambda\mu} \tilde{a}^\lambda \tilde{b}^\mu \tag{1.13}$$

for any transformation $x^\mu \to \tilde{x}^\mu$ and arbitrary vectors a^λ and b^μ. We can define also a mixed tensor $T^\lambda{}_\mu$ of order two by the demand:

$$T^\lambda{}_\mu a_\lambda b^\mu = \tilde{T}^\lambda{}_\mu \tilde{a}_\lambda \tilde{b}^\mu , \tag{1.14}$$

again for any $x^\mu \to \tilde{x}^\mu$ and arbitrary vectors a_λ and b^μ.

The transformation formulae for the components of the tensors of order 2 follow from the definitions (1.12) to (1.14) combined with the relations (1.4) and (1.10′). The results are:

$$\left.\begin{aligned} \tilde{T}^{\lambda\mu} &= \frac{\partial \tilde{x}^\lambda}{\partial x^\alpha} \frac{\partial \tilde{x}^\mu}{\partial x^\beta} T^{\alpha\beta} ; \qquad \tilde{T}_{\lambda\mu} = \frac{\partial x^\alpha}{\partial \tilde{x}^\lambda} \frac{\partial x^\beta}{\partial \tilde{x}^\mu} T_{\alpha\beta} ; \\ \tilde{T}^\lambda{}_\mu &= \frac{\partial \tilde{x}^\lambda}{\partial x^\alpha} \frac{\partial x^\beta}{\partial \tilde{x}^\mu} T^\alpha{}_\beta . \end{aligned}\right\} \tag{1.15}$$

As an application of the last of Equations (1.15) let us consider the mixed tensor which has in the frame x^μ the components $T^\alpha{}_\beta = \delta^\alpha{}_\beta$. We then derive at once from (1.15) and (1.6):

$$\tilde{T}^\lambda{}_\mu = \delta^\lambda_\mu = T^\lambda{}_\mu .$$

Thus the Kronecker symbol δ^λ_μ is a mixed tensor having frame-independent values for its components.

In the most general case we shall have a tensor $T^{\lambda_1\lambda_2\ldots}{}_{\mu_1\mu_2\ldots}$ of order (p, q), $p \geqslant 0$ being the number of upper indices and $q \geqslant 0$ of lower indices. The tensor property will be equivalent to the following requirement. : If $a_{\lambda_1}, a_{\lambda_2}, \ldots, a_{\lambda_p}$ are p arbitrary covariant vectors and $b^{\mu_1}, \ldots, b^{\mu_q}$ are q arbitrary contravariant vectors, the sum $T^{\lambda_1\lambda_1\ldots}{}_{\mu_1\mu_2\ldots} \times \times a_{\lambda_1} a_{\lambda_2} \ldots b^{\mu_1} b^{\mu_2} \ldots$ must be a scalar. The transformation formula for the tensor $T^{\lambda_1\ldots}{}_{\mu_1\ldots}$ can be derived immediately from this requirement.

Note that a tensor of order (0, 0) is a scalar. A tensor of order (1, 0) is a contravariant vector and of order (0, 1) a covariant vector.

2. Algebraic Operations. Symmetry Properties

Tensors of the same order (p, q) can be added, their *sum* being again a tensor of the same order. For example, if we have two contravariant vectors a^λ and b^λ, we find from the transformation formula (1.4) applied to each one of them:

$$\tilde{a}^\lambda + \tilde{b}^\lambda = \frac{\partial \tilde{x}^\lambda}{\partial x^\alpha}(a^\alpha + b^\alpha). \tag{2.1}$$

This relation means, according to (1.4), that $a^\lambda + b^\lambda$ is a contravariant vector. Note that the tensors, which are to be added, must be given at the same point of the space. This is so because we allow general coordinate transformations and consequently $\partial \tilde{x}^\lambda / \partial x^\alpha$ will in general change from one point to another.

The *product* of two vectors is a tensor of order 2. Indeed, we derive from (1.4) at once the relation:

$$\tilde{a}^\lambda \tilde{b}^\mu = \frac{\partial \tilde{x}^\lambda}{\partial x^\alpha} \frac{\partial \tilde{x}^\mu}{\partial x^\beta} a^\alpha b^\beta, \tag{2.2}$$

which shows that $a^\lambda b^\mu \equiv T^{\lambda\mu}$ is a tensor of order (2, 0). In a similar way we can multiply tensors of any order. One sees at once that if the two factors are respectively of order (p, q) and (p', q'), the product will be a tensor of order $(p+p', q+q')$. The extension to more than two factors is straightforward.

A *contraction* is possible for any tensor of order (p, q) with p and $q > 0$. Consider for example the tensor $T^{\lambda\mu}{}_\nu$. Putting $\nu = \mu$ we get the expression $T^{\lambda\mu}{}_\mu$ which, according to the summation convention (1.11), is a sum over the values of the index μ. Using the transformation formula for $T^{\lambda\mu}{}_\nu$,

$$\tilde{T}^{\lambda\mu}{}_\nu = \frac{\partial \tilde{x}^\lambda}{\partial x^\alpha} \frac{\partial \tilde{x}^\mu}{\partial x^\beta} \frac{\partial x^\gamma}{\partial \tilde{x}^\nu} T^{\alpha\beta}{}_\gamma,$$

and the relations (1.6) we find at once:

$$\tilde{T}^{\lambda\mu}{}_\mu = \frac{\partial \tilde{x}^\lambda}{\partial x^\alpha} T^{\alpha\beta}{}_\beta. \tag{2.3}$$

This result shows that $T^{\lambda\mu}{}_\mu$ is a contravariant vector. In the general case of a tensor of order (p, q) the result of the contraction will be a tensor of order $(p-1, q-1)$.

We can combine multiplication and contraction. E.g. starting from the vectors a^λ and b_μ we first form the mixed tensor of order (1, 1):

$$T^\lambda{}_\mu = a^\lambda b_\mu,$$

and then the scalar

$$T^{\lambda}{}_{\lambda} = a^{\lambda} b_{\lambda} .$$

The scalar $T^{\lambda}{}_{\lambda}$ is called the *trace* of the mixed tensor $T^{\lambda}{}_{\mu}$.

In order to discuss symmetry properties of tensors we consider first a contravariant tensor $T^{\lambda\mu}$. The components of such a tensor can be written conveniently in the form of a matrix:

$$T^{\lambda\mu} = \begin{pmatrix} T^{11} & T^{12} & \dots & T^{1n} \\ T^{21} & T^{22} & \dots & T^{2n} \\ \dots & \dots & \dots & \dots \\ T^{n1} & T^{n2} & \dots & T^{nn} \end{pmatrix} .$$

Suppose now that in a given frame x^{μ} this matrix is symmetric:

$$T^{\lambda\mu} = T^{\mu\lambda} . \tag{2.4}$$

Then we see at once from the transformation formula (1.15) that the matrix representing the components of this tensor will be symmetric in *all* frames. We say in this case that the tensor $T^{\lambda\mu}$ is *symmetric*.

Similarly, if the matrix $T^{\lambda\mu}$ is antisymmetric in x^{μ}:

$$T^{\lambda\mu} = - T^{\mu\lambda} , \tag{2.5}$$

then it will be antisymmetric in all frames. The tensor $T^{\lambda\mu}$ is called in this case *antisymmetric*.

The same remarks are valid for a covariant tensor $T_{\lambda\mu}$. Again, if $T_{\lambda\mu} = T_{\mu\lambda}$, the tensor will be called symmetric; if $T_{\lambda\mu} = -T_{\mu\lambda}$, the tensor is antisymmetric.

No symmetry properties can be defined for a mixed tensor $T^{\lambda}{}_{\mu}$. The matrix of the components of $T^{\lambda}{}_{\mu}$ could be symmetric in some frame x^{μ}, but this property would not be conserved in a coordinate transformation.

A contravariant or covariant tensor of order 2 can be written as the sum of a symmetric and an antisymmetric tensor:

$$T_{\lambda\mu} = S_{\lambda\mu} + A_{\lambda\mu} ; \qquad S_{\lambda\mu} = S_{\mu\lambda} , \qquad A_{\mu\lambda} = - A_{\lambda\mu} . \tag{2.6}$$

The tensors $S_{\lambda\mu}$ and $A_{\lambda\mu}$ are determined uniquely:

$$S_{\lambda\mu} = \tfrac{1}{2}(T_{\lambda\mu} + T_{\mu\lambda}), \qquad A_{\lambda\mu} = \tfrac{1}{2}(T_{\lambda\mu} - T_{\mu\lambda}) . \tag{2.7}$$

It will be usefull to introduce the following notation:

$$T_{(\lambda\mu)} = \tfrac{1}{2}(T_{\lambda\mu} + T_{\mu\lambda}) = S_{\lambda\mu} , \qquad T_{[\lambda\mu]} = \tfrac{1}{2}(T_{\lambda\mu} - T_{\mu\lambda}) = A_{\lambda\mu} . \tag{2.8}$$

The number of independent components of the symmetric tensor $T_{(\lambda\mu)}$ is found easily to be

$$n + \frac{n^2 - n}{2} = \frac{n(n+1)}{2} .$$

The antisymmetric tensor $T_{[\lambda\mu]}$ has $(n^2-n)/2=[n(n-1)]/2$ independent components. This is an agreement with the fact that the tensor $T_{\lambda\mu}$ has n^2 components, since

$$\frac{n(n+1)}{2}+\frac{n(n-1)}{2}=n^2.$$

In the case of a tensor of order (p, q) we can define symmetry or antisymmetry with respect to a pair of indices which are both upper or lower. E.g. the tensor $T^{\lambda\mu\nu}$ will be called symmetric or antisymmetric in λ, μ if

$$T^{\lambda\mu\nu}=T^{\mu\lambda\nu} \quad \text{or} \quad T^{\lambda\mu\nu}=-T^{\mu\lambda\nu}.$$

This property is conserved in a coordinate transformation.

Especially interesting is the case of a tensor having only upper or only lower indices. Such a tensor will be called totally symmetric or simply *symmetric* if it is symmetric with respect to any pair of indices. E.g. the tensor $T^{\lambda\mu\nu}$ is symmetric if

$$T^{\lambda\mu\nu}=T^{\mu\lambda\nu}=T^{\lambda\nu\mu}=T^{\nu\mu\lambda}.$$

Similarly, the tensor $T^{\lambda\mu\nu}$ will be called totally antisymmetric or simply *antisymmetric* if

$$T^{\lambda\mu\nu}=-T^{\mu\lambda\nu}=-T^{\lambda\nu\mu}=-T^{\nu\mu\lambda}.$$

The notation introduced in (2.8) can be generalised to tensors of higher order. For a tensor $T_{\lambda\mu\nu}$ of order (0, 3) we write:

$$T_{(\lambda\mu\nu)}=\tfrac{1}{6}(T_{\lambda\mu\nu}+T_{\mu\nu\lambda}+T_{\nu\lambda\mu}+T_{\nu\mu\lambda}+T_{\mu\lambda\nu}+T_{\lambda\nu\mu}), \tag{2.9}$$

$$T_{[\lambda\mu\nu]}=\tfrac{1}{6}(T_{\lambda\mu\nu}+T_{\mu\nu\lambda}+T_{\nu\lambda\mu}-T_{\nu\mu\lambda}-T_{\mu\lambda\nu}-T_{\lambda\nu\mu}). \tag{2.10}$$

It is verified at once that $T_{(\lambda\mu\nu)}$ is symmetric and $T_{[\lambda\mu\nu]}$ antisymmetric. The same notation can be used for a contravariant tensor.

With the notation (2.9) and (2.10) we can write the condition for a tensor $T_{\lambda\mu\nu}$ to be symmetric or antisymmetric in the form:

$$T_{\lambda\mu\nu}=T_{(\lambda\mu\nu)} \quad \text{or} \quad T_{\lambda\mu\nu}=T_{[\lambda\mu\nu]}.$$

The case of antisymmetric tensors deserves further discussion. We first remark that according to the definition (2.5) a contravariant tensor $T^{\lambda\mu\cdots}$, which is antisymmetric in λ, μ, can have non-vanishing components only when $\lambda\neq\mu$. Consequently an antisymmetric tensor $T^{\lambda\mu\nu\cdots}$ can have non-vanishing components only for values of the indices λ, μ, ν, ... which are all different. It follows at once that all components of an antisymmetric tensor of order $(p, 0)$, $p>n$ are identically zero. I.e. the highest order of an antisymmetric tensor is $p=n$.

Consider as an example the case $n=4$. The antisymmetric tensor of highest order is a tensor $T^{\lambda\mu\nu\varrho}$ of order $p=4$. The non-vanishing components of this tensor are those for which λ, μ, ν and ϱ are permutations of the numbers 1, 2, 3 and 4. Moreover we have the relation

$$T^{\lambda\mu\nu\varrho}=\pm T^{1234}, \tag{2.11}$$

the positive sign corresponding to the $\lambda\mu\nu\varrho$ being an even permutation of 1234 and the negative sign to the opposite case. We thus see that an antisymmetric tensor of order $p=n$ has only one component, resembling in this respect a scalar. This component, however, does depend on the coordinate system in a manner which we shall determine in Section 3. For this reason such a tensor is often called a pseudoscalar.

It can be seen easily that a contravariant antisymmetric tensor of order $p=n-1$ has n independent components, i.e. as many as a vector; thus this tensor is often called a pseudovector. An antisymmetric tensor of order $p=n-2$ has as many independent components as an antisymmetric tensor of order $p=2$ and so on. All of these remarks are valid also for covariant antisymmetric tensors.

To end this chapter we give a construction of an antisymmetric tensor of order $p \leqslant n$ from p independent vectors. It will be sufficient to write the formula for $p=3$. Let the 3 vectors a^λ, b^λ and c^λ be linearly independent. Then an antisymmetric tensor of order $p=3$ is defined by the formula:

$$T^{\lambda\mu\nu} = \begin{vmatrix} a^\lambda & b^\lambda & c^\lambda \\ a^\mu & b^\mu & c^\mu \\ a^\nu & b^\nu & c^\nu \end{vmatrix}, \tag{2.12}$$

the antisymmetry being an immediate consequence of the properties of determinants. This type of tensor will be used later in order to define the volumelement of an n-dimensional space.

3. Tensor Densities

Let us consider a covariant antisymmetric tensor of order $q=n$. In the example $n=4$ the non-vanishing components of this tensor $T_{\lambda\mu\nu\varrho}$ are equal to $\pm T_{1234}$. We now ask the question: How does T_{1234} change under a coordinate transformation?

The answer will be found from the transformation formula,

$$\tilde{T}_{1234} = \frac{\partial x^\lambda}{\partial \tilde{x}^1}\frac{\partial x^\mu}{\partial \tilde{x}^2}\frac{\partial x^\nu}{\partial \tilde{x}^3}\frac{\partial x^\varrho}{\partial \tilde{x}^4} T_{\lambda\mu\nu\varrho}.$$

The sum in the right-hand side is over all non-vanishing components $T_{\lambda\mu\nu\varrho}$. Their number is $4!=24$, 12 of them being equal to T_{1234} (when $\lambda\mu\nu\varrho$ is an even permutation of 1234) and the other 12 equal to $-T_{1234}$ (odd permutations of 1234). The final result of the summation is extremely simple:

$$\tilde{T}_{1234} = D \cdot T_{1234}, \tag{3.1}$$

$$D = \det \frac{\partial x^\alpha}{\partial \tilde{x}^\lambda}. \tag{3.2}$$

More generally we have for any n:

$$\tilde{T}_{12\ldots n} = D \cdot T_{12\ldots n}. \tag{3.1'}$$

Quantities with one component Q transforming like (3.1′),

$$\tilde{Q} = D \cdot Q, \tag{3.3}$$

can be constructed also in other ways. More generally there are quantities with one component Q transforming as follows:

$$\tilde{Q} = D^w \cdot Q, \tag{3.4}$$

w being a positive or negative integer. A quantity Q transforming like (3.4) will be called a *scalar density of weight* w.

An example of such a density of weight $w=2$ is obtained as follows. Start from the transformation formula (1.15) for a covariant tensor $T_{\lambda\mu}$. This relation can be considered as a matrix equation. It follows then at once from the rule for the determinant of a product of matrices that

$$\det \tilde{T}_{\lambda\mu} = D^2 \cdot \det T_{\lambda\mu} . \tag{3.5}$$

Therefore the determinant of a covariant tensor of order 2 is a scalar density of weight $w=2$.

The product of a tensor of order (p, q) by a scalar density of weight w is by definition a *tensor-density* of order (p, q) and weight w. The transformation formula for a tensor density follows at once from the corresponding formulae for tensors and scalar densities. E.g. for a covariant vector-density of weight $w=1$, $\mathfrak{A}_\lambda = A_\lambda Q$:

$$\tilde{\mathfrak{A}}_\lambda = \tilde{A}_\lambda \tilde{Q} = \frac{\partial x^\alpha}{\partial \tilde{x}^\lambda} A_\alpha \cdot QD = \frac{\partial x^\alpha}{\partial \tilde{x}^\lambda} \mathfrak{A}_\alpha D . \tag{3.6}$$

More generally the transformation formula for a tensor-density of weight w will differ from the formula for the corresponding tensor by a factor D^w appearing on the right-hand side of the equation.

The preceding definition of a tensor-density leads immediately to the following results:

(i) We can add tensor-densities, given at the same point, if they are of the same order (p, q) and the same weight w.

(ii) We can always multiply tensor densities (given at the same point). The product of a tensor-density of order (p, q) and weight w by another of order (p', q') and weight w' is a tensor-density of order $(p+p', q+q')$ and weight $w+w'$.

(iii) A tensor-density of order (p, q) and weight w, with p and $q>0$, can be contracted. The result of the contraction is a tensor-density of order $(p-1, q-1)$ and of weight w.

The *symmetry properties* for tensor densities are defined exactly as for tensors. Indeed, a tensor-density $\mathfrak{A}^{\lambda\mu\ldots}{}_{\varrho\sigma\ldots}$ can be written as the product of a tensor by a scalar density:

$$\mathfrak{A}^{\lambda\mu..}{}_{\varrho\sigma..} = A^{\lambda\mu..}{}_{\varrho\sigma..} Q . \tag{3.7}$$

We ascribe to the tensor-density $\mathfrak{A}^{\lambda\mu..}{}_{\varrho\sigma..}$ the symmetry properties of the tensor $A^{\lambda\mu\ldots}{}_{\varrho\sigma\ldots}$.

As an application of the preceding definitions let us consider an antisymmetric tensor-density of order $(n, 0)$ and weight $w=1$. In order to simplify the formulae we take $n=4$. The non-vanishing components of this tensor-density $\varepsilon^{\lambda\mu\nu\varrho}$ are equal to $\pm\varepsilon^{1234}$. To define this tensor-density completely we assume that in a given coordinate system x^μ we have

$$\varepsilon^{1234} = 1\,. \tag{3.8}$$

What will be the value of $\tilde{\varepsilon}^{1234}$ in some other coordinate system $\tilde{x}^\mu$?

From the transformation formula

$$\tilde{\varepsilon}^{1234} = \frac{\partial\tilde{x}^1}{\partial x^\alpha}\frac{\partial\tilde{x}^2}{\partial x^\beta}\frac{\partial\tilde{x}^3}{\partial x^\gamma}\frac{\partial\tilde{x}^4}{\partial x^\delta}\,\varepsilon^{\alpha\beta\gamma\delta}\cdot D$$

we find at once:

$$\tilde{\varepsilon}^{1234} = \det\frac{\partial\tilde{x}^\lambda}{\partial x^\alpha}\cdot\varepsilon^{1234}D = \det\frac{\partial\tilde{x}^\lambda}{\partial x^\alpha}\cdot D\,.$$

Now from Equation (1.6) considered as a matrix equation we derive the relation:

$$\det\frac{\partial\tilde{x}^\lambda}{\partial x^\alpha}\cdot\det\frac{\partial x^\beta}{\partial\tilde{x}^\mu} = \det\frac{\partial\tilde{x}^\lambda}{\partial x^\alpha}\cdot D = 1\,.$$

We have therefore

$$\tilde{\varepsilon}^{1234} = \varepsilon^{1234} = 1\,. \tag{3.9}$$

I.e. the components $\varepsilon^{\lambda\mu\nu\varrho}$ do not depend on the coordinate system. This is the *Levi-Civita tensor-density*, which is extremely useful in many calculations.

As an example of the use of this tensor-density let us consider in a space with $n=4$ the antisymmetric tensors $A_{\lambda\mu\nu\varrho}$ and $A_{\lambda\mu\nu}$. Multiplication of these tensors by $\varepsilon^{\lambda\mu\nu\varrho}$ with repeated contraction gives:

$$\left.\begin{aligned}
&\varepsilon^{\lambda\mu\nu\varrho}A_{\lambda\mu\nu\varrho} = 4!\,A_{1234} = \text{scalar density of weight } 1\,;\\
&\varepsilon^{\lambda\mu\nu\varrho}A_{\lambda\mu\nu} \equiv \mathfrak{A}^\varrho = \text{contravariant vector density of weight } 1\,,
\end{aligned}\right\} \tag{3.10}$$

$$\tfrac{1}{6}\mathfrak{A}^\varrho = (-A_{234}, -A_{314}, -A_{124}, A_{123})\,. \tag{3.11}$$

Thus we see that what we called in Section 2 a pseudoscalar or a pseudovector is more exactly a scalar density or a vector-density.

The concept of an antisymmetric tensor-density of order $p=n$ can be used in order to define the volumelement in a general space of n dimensions. Let us first consider a 3-dimensional euclidean space. When we use orthogonal cartesian coordinates, the volume of the parallelepiped formed by the 3 infinitesimal displacement vectors ${}_1\mathrm{d}x^\lambda$, ${}_2\mathrm{d}x^\lambda$ and ${}_3\mathrm{d}x^\lambda$ is given by the determinant

$$\begin{vmatrix} {}_1\mathrm{d}x^1 & {}_2\mathrm{d}x^1 & {}_3\mathrm{d}x^1 \\ {}_1\mathrm{d}x^2 & {}_2\mathrm{d}x^2 & {}_3\mathrm{d}x^2 \\ {}_1\mathrm{d}x^3 & {}_2\mathrm{d}x^3 & {}_3\mathrm{d}x^3 \end{vmatrix} = \delta V\,. \tag{3.12}$$

This is the component δA^{123} of the contravariant antisymmetric tensor $\delta A^{\lambda\mu\nu}$ which is constructed from the vectors ${}_1\mathrm{d}x^\lambda$, ${}_2\mathrm{d}x^\lambda$ and ${}_3\mathrm{d}x^\lambda$ according to the formula (2.12).

In another coordinate system $\tilde{x}^\mu$ we shall obtain $\delta \tilde{A}^{123}$ from the transformation formula for $\delta A^{\lambda\mu\nu}$. A calculation similar to that which led to the relation (3.1) leads in the present case – in which we have a *contravariant* antisymmetric tensor of order $p=n$ – to the result:

$$\delta \tilde{A}^{123} = \frac{1}{D}\, \delta A^{123}. \tag{3.13}$$

Therefore:

$$\delta \tilde{V} = \frac{1}{D}\, \delta V. \tag{3.14}$$

If the coordinates $\tilde{x}^\mu$ are also orthogonal cartesian, with the axes oriented in the same way as in x^μ, then $D=1$: For such transformations we shall have $\delta\tilde{V}=\delta V$.

In a Euclidean space with non-Cartesian coordinates, or in a curved space, in which no Cartesian coordinates exist, we have only the relation (3.14), with $D\neq 1$ in general. For this reason we call δV the *non-invariant* volume element of the 3-dimensional space.

If we have a scalar density Q of weight $w=1$ defined in a region V of the space, we can construct a scalar by multiplying δV by Q:

$$Q\delta V = \tilde{Q}\delta\tilde{V} = \text{scalar}. \tag{3.15}$$

We now remark that scalars can be added also when they are defined at different points of the space. We can therefore define a coordinate-independent or scalar integral over the region V:

$$\int_V Q\, \delta V = \text{scalar}. \tag{3.16}$$

The generalisation of these definitions for a space of dimensions $n\neq 3$ is straightforward. E.g. in the case $n=4$ the only difference will be that we have to start with 4 independent vectors ${}_a\mathrm{d}x^\lambda$ $(a=1, 2, 3, 4)$ in order to build the non-invariant volume element

$$\delta V = \delta A^{1234}. \tag{3.17}$$

We shall see later that in a Riemannian space there is a covariant tensor of order 2 having a geometrical meaning: the metrical tensor $g_{\mu\nu}$. The quantity $(\det g_{\mu\nu})^{1/2}$ is then, according to (3.5), a scalar density of weight 1 and consequently the product $(\det g_{\mu\nu})^{1/2}\, \delta V$ is a scalar. This will be the invariant volume element of the Riemannian space. In a general, non-metrical space we have only the non-invariant volume element δV.

Exercises

I1: Verify the transformation formulae (1.15).

I2: Derive the transformation formula for a tensor $T^{\lambda\mu\cdots}{}_{\varrho\sigma\cdots}$ of order (p, q).

I3: Prove that an antisymmetric tensor density of order $(0, n)$ and weight -1 has invariant components.

CHAPTER II

COVARIANT DIFFERENTIATION

4. Differentiation

We consider a region V of the space in which some tensor, e.g. a covariant vector a_λ, is given at each point $P(x^\alpha)$:

$$a_\lambda = a_\lambda(x^\alpha). \tag{4.1}$$

We say then that we are given a *tensor field* in V. We shall assume that the components of the tensor are continuous and differentiable functions of x^α and we ask the question: Is it possible to construct a new tensor field by differentiating the given one?

We start with the simplest tensor field which is the one of order zero, i.e. a scalar field $\varphi = \varphi(x^\alpha)$. The derivatives

$$\frac{\partial \varphi}{\partial x^\alpha} \equiv \varphi_{,\alpha} \tag{4.2}$$

are seen at once to be the components of a covariant vector. This follows from the relation

$$\mathrm{d}\varphi = \varphi_{,\alpha}\,\mathrm{d}x^\alpha, \tag{4.3}$$

valid for arbitrary $\mathrm{d}x^\alpha$. Since $\mathrm{d}\varphi$ is a scalar and $\mathrm{d}x^\alpha$ an arbitrary contravariant vector, $\varphi_{,\alpha}$ will be a covariant vector. The same result follows also from the elementary formula

$$\frac{\partial \varphi}{\partial \tilde{x}^\lambda} = \frac{\partial \varphi}{\partial x^\alpha}\frac{\partial x^\alpha}{\partial \tilde{x}^\lambda}, \tag{4.3'}$$

which is exactly the transformation formula (1.10′) for covariant vectors. The vector-field $\varphi_{,\alpha}$ is the *gradient* of the scalar field φ.

Next we consider a covariant vector field $a_\lambda(x^\alpha)$. We derive the relation between

$$\frac{\partial a_\lambda}{\partial x^\mu} \equiv a_{\lambda,\mu} \tag{4.4}$$

and the corresponding expression $\tilde{a}_{\lambda,\mu}$ in some other coordinate system $\tilde{x}^\mu$:

$$\tilde{a}_{\lambda,\mu} \equiv \frac{\partial \tilde{a}_\lambda}{\partial \tilde{x}^\mu} = \frac{\partial}{\partial \tilde{x}^\mu}\left(\frac{\partial x^\varrho}{\partial \tilde{x}^\lambda}\, a_\varrho\right) = \frac{\partial^2 x^\varrho}{\partial \tilde{x}^\lambda \partial \tilde{x}^\mu}\, a_\varrho + \frac{\partial x^\varrho}{\partial \tilde{x}^\lambda}\frac{\partial x^\sigma}{\partial \tilde{x}^\mu}\, a_{\varrho,\sigma}\,. \tag{4.5}$$

Without the first term in the right hand side this equation would be the transformation formula for a covariant tensor of order 2. This term would vanish in the case of a linear transformation, e.g. a Lorentz transformation in Minkowski space. Since we are here interested in the case of arbitrary coordinates as well as transformations, $a_{\lambda,\mu}$ is not a tensor.

The general method for overcoming the difficulty represented by the first term in the right hand side of (4.5) is based on the introduction of a *connection*, as we shall see in Section 5. Here we shall describe some special, but interesting cases in which a connection is not needed. In all of these cases we shall consider antisymmetric covariant tensors or antisymmetric contravariant tensor-densities of weight 1.

Noting that the perturbing term in (4.5) is symmetric in λ, μ we find at once:

$$\tilde{a}_{\lambda,\mu} - \tilde{a}_{\mu,\lambda} = \frac{\partial x^{\varrho}}{\partial \tilde{x}^{\lambda}} \frac{\partial x^{\sigma}}{\partial \tilde{x}^{\mu}} (a_{\varrho,\sigma} - a_{\sigma,\varrho}). \tag{4.6}$$

Thus the difference

$$a_{\varrho,\sigma} - a_{\sigma,\varrho} \equiv 2a_{[\varrho,\sigma]}$$

is a covariant antisymmetric tensor. This is the *rotation* of the vector a_{λ}, generalising the curl of the 3-dimensional space.

We now start from an antisymmetric tensor $a_{\lambda\mu} = -a_{\mu\lambda}$. We derive the relation between $a_{\lambda\mu,\nu}$ and the corresponding quantity $\tilde{a}_{\lambda\mu,\nu}$ in some other coordinate system $\tilde{x}^{\mu}$. Differentiating the transformation formula (1.15) for a tensor $a_{\lambda\mu}$ we find:

$$\tilde{a}_{\lambda\mu,\nu} = \left(\frac{\partial^2 x^{\alpha}}{\partial \tilde{x}^{\nu} \partial \tilde{x}^{\lambda}} \frac{\partial x^{\beta}}{\partial \tilde{x}^{\mu}} + \frac{\partial x^{\alpha}}{\partial \tilde{x}^{\lambda}} \frac{\partial^2 x^{\beta}}{\partial \tilde{x}^{\mu} \partial \tilde{x}^{\nu}} \right) a_{\alpha\beta} + \frac{\partial x^{\alpha}}{\partial \tilde{x}^{\lambda}} \frac{\partial x^{\beta}}{\partial \tilde{x}^{\mu}} \frac{\partial x^{\gamma}}{\partial \tilde{x}^{\nu}} a_{\alpha\beta,\gamma}. \tag{4.7}$$

If we now take the antisymmetric part of this relation according to Equation (2.10), we find that the terms containing second derivatives cancel in the summation and the final result is:

$$\tilde{a}_{[\lambda\mu,\nu]} = \frac{\partial x^{\alpha}}{\partial \tilde{x}^{\lambda}} \frac{\partial x^{\beta}}{\partial \tilde{x}^{\mu}} \frac{\partial x^{\gamma}}{\partial \tilde{x}^{\nu}} a_{[\alpha\beta,\gamma]}, \tag{4.8}$$

which means that $a_{[\alpha\beta,\gamma]}$ is a tensor. Note that because of the antisymmetry of $a_{\alpha\beta,\gamma}$ with respect to α, β the general expression (2.10) reduces in this case to 3 terms:

$$a_{[\alpha\beta,\gamma]} = \tfrac{1}{3}(a_{\alpha\beta,\gamma} + a_{\beta\gamma,\alpha} + a_{\gamma\alpha,\beta}). \tag{4.9}$$

Starting from an antisymmetric covariant tensor $a_{\lambda\mu\nu} = a_{[\lambda\mu\nu]}$ we prove in a similar way that the quantity $a_{[\lambda\mu\nu,\varrho]}$ is an antisymmetric tensor of order 4. Because of the antisymmetry of $a_{\lambda\mu\nu}$ the expression for $a_{[\lambda\mu\nu,\varrho]}$, which in the general case would contain $4! = 24$ terms, reduces to 4 terms only:

$$a_{[\lambda\mu\nu,\varrho]} = \tfrac{1}{4}(a_{\lambda\mu\nu,\varrho} - a_{\mu\nu\varrho,\lambda} + a_{\nu\varrho\lambda,\mu} - a_{\varrho\lambda\mu,\nu}). \tag{4.10}$$

There is another type of differential operation applied on antisymmetric contravariant tensor densities of weight 1, in which there is no need of a connection. They can be derived very easily from the preceding formulae established for covariant antisymmetric tensors with the help of the Levi-Civita tensor-density $\varepsilon^{\lambda\mu\nu\varrho}$. To avoid lengthy equations we consider the case $n=4$. The generalisation for any n is straightforward.

Multiplying (4.10) by $\varepsilon^{\lambda\mu\nu\varrho}$ we find:

$$\varepsilon^{\lambda\mu\nu\varrho} a_{[\lambda\mu\nu,\varrho]} = \varepsilon^{\lambda\mu\nu\varrho} a_{\lambda\mu\nu,\varrho} .$$

Since $\varepsilon^{\lambda\mu\nu\varrho}$ is constant, we can write this relation in the form

$$\varepsilon^{\lambda\mu\nu\varrho} a_{[\lambda\mu\nu,\varrho]} = (\varepsilon^{\lambda\mu\nu\varrho} a_{\lambda\mu\nu})_{,\varrho} . \tag{4.11}$$

The left-hand side is evidently a scalar density, while $\varepsilon^{\lambda\mu\nu\varrho} a_{\lambda\mu\nu}$ is a vector-density (both of weight $w=1$):

$$\varepsilon^{\lambda\mu\nu\varrho} a_{[\lambda\mu\nu,\varrho]} = \mathfrak{B} ; \qquad \varepsilon^{\lambda\mu\nu\varrho} a_{\lambda\mu\nu} = \mathfrak{A}^{\varrho} .$$

Therefore the meaning of Equation (4.11),

$$\mathfrak{A}^{\varrho}{}_{,\varrho} = \mathfrak{B} , \tag{4.12}$$

is that the divergence of a contravariant vector-density is a scalar density.

If we multiply Equation (4.9) by $\varepsilon^{\lambda\mu\nu\varrho}$ we find:

$$\varepsilon^{\lambda\mu\nu\varrho} a_{[\lambda\mu,\nu]} = (\varepsilon^{\lambda\mu\nu\varrho} a_{\lambda\mu})_{,\nu} . \tag{4.13}$$

The expression on the left-hand side is evidently a vector-density $\mathfrak{B}^{\varrho}$, while $\varepsilon^{\lambda\mu\nu\varrho} a_{\lambda\mu}$ is an antisymmetric tensor-density $\mathfrak{A}^{\nu\varrho}$. Thus we have:

$$\mathfrak{A}^{\nu\varrho}{}_{,\nu} = \mathfrak{B}^{\varrho} : \tag{4.13'}$$

The 'divergence' of an antisymmetric tensor-density $\mathfrak{A}^{\nu\varrho}$ is a contravariant vector density.

Similarly we find the relations:

$$\left.\begin{aligned} \mathfrak{A}^{\mu\nu\varrho}{}_{,\mu} &= \mathfrak{B}^{\nu\varrho} ; \\ \mathfrak{A}^{\lambda\mu\nu\varrho}{}_{,\lambda} &= \mathfrak{B}^{\mu\nu\varrho} . \end{aligned}\right\} \tag{4.14}$$

All tensor densities entering in these relations are antisymmetric and of weight $w=1$. Note that the second line of (4.14) is the relation of the highest order which we can construct in a 4-dimensional space.

As an application of the preceding formulae let us consider Maxwell's equations, which in an inertial frame of Minkowski space are:

$$\eta^{\mu\nu} F_{\lambda\mu,\nu} = s_{\lambda} ; \qquad F_{[\lambda\mu,\nu]} = 0 .$$

Since $F_{\lambda\mu}$ is antisymmetric, it follows from (4.8) that the second of these equations has already its general form. The first equation has to be written in the form (4.13'):

$$\mathfrak{F}^{\lambda\mu}{}_{,\mu} = \mathfrak{s}^{\lambda} ,$$

$\mathfrak{F}^{\lambda\mu}$ and $\mathfrak{s}^{\lambda}$ being now tensor densities of weight $w=1$. In a Riemannian space the tensor density $\mathfrak{F}^{\lambda\mu}$ is derived from the tensor $F_{\lambda\mu}$ with the help of the metrical tensor. We shall derive this relation in Section 36.

5. The Connection

We start with a heuristic argument. Returning to the relation (4.5) we rewrite it with the help of (1.10) in the form:

$$\tilde{a}_{\lambda,\mu} - \frac{\partial^2 x^\varrho}{\partial\tilde{x}^\lambda \partial\tilde{x}^\mu}\frac{\partial\tilde{x}^\nu}{\partial x^\varrho}\tilde{a}_\nu = \frac{\partial x^\varrho}{\partial\tilde{x}^\lambda}\frac{\partial x^\sigma}{\partial\tilde{x}^\mu}a_{\varrho,\sigma}. \tag{5.1}$$

The components of a tensor in a given frame can be chosen arbitrarily. Let us assume that there is a tensor $A_{\lambda\mu}$ whose components in x^μ are the derivatives $a_{\lambda,\mu}$. The meaning of the relation (5.1) is then the following: According to the second of Equations (1.15), the components $\tilde{A}_{\lambda\mu}$ of the tensor $A_{\lambda\mu}$ in $\tilde{x}^\mu$ are:

$$\tilde{A}_{\lambda\mu} = \tilde{a}_{\lambda,\mu} - \tilde{\gamma}^\nu_{\lambda\mu}\tilde{a}_\nu\,; \tag{5.2}$$

$$\tilde{\gamma}^\nu_{\lambda\mu} = \frac{\partial\tilde{x}^\nu}{\partial x^\varrho}\frac{\partial^2 x^\varrho}{\partial\tilde{x}^\lambda\partial\tilde{x}^\mu}. \tag{5.2'}$$

This result shows that, while the derivatives $a_{\lambda,\mu}$ could be the components of a tensor in some special frame, we have to try in the general case to construct a tensor by using a new coordinate-dependent quantity $\Gamma^\varrho_{\lambda\mu}$:

$$a_{\lambda,\mu} - \Gamma^\varrho_{\lambda\mu}a_\varrho \equiv A_{\lambda\mu} = \text{a tensor}. \tag{5.3}$$

In another coordinate system $\tilde{x}^\mu$ the components $\tilde{A}_{\lambda\mu}$ will then be:

$$\tilde{a}_{\lambda,\mu} - \tilde{\Gamma}^\varrho_{\lambda\mu}\tilde{a}_\varrho = \tilde{A}_{\lambda\mu}. \tag{5.4}$$

The transformation formula for a covariant tensor $A_{\lambda\mu}$ will allow us to determine the transformation properties of the quantity $\Gamma^\varrho_{\lambda\mu}$. Indeed, we find from the second of Equations (1.15):

$$\tilde{a}_{\lambda,\mu} - \tilde{\Gamma}^\varrho_{\lambda\mu}\tilde{a}_\varrho = \frac{\partial x^\alpha}{\partial\tilde{x}^\lambda}\frac{\partial x^\beta}{\partial\tilde{x}^\mu}(a_{\alpha,\beta} - \Gamma^\gamma_{\alpha\beta}a_\gamma).$$

Taking into account Equation (5.1) and expressing the factor a_γ in the last term with the help of Equation (1.10) we rewrite this relation in the form:

$$0 = \tilde{a}_\alpha\left\{\tilde{\Gamma}^\alpha_{\lambda\mu} - \frac{\partial x^\beta}{\partial\tilde{x}^\lambda}\frac{\partial x^\gamma}{\partial\tilde{x}^\mu}\frac{\partial\tilde{x}^\alpha}{\partial x^\varrho}\Gamma^\varrho_{\beta\gamma} - \frac{\partial\tilde{x}^\alpha}{\partial x^\beta}\frac{\partial^2 x^\beta}{\partial\tilde{x}^\lambda\partial\tilde{x}^\mu}\right\}.$$

Since this must be valid for any vector $\tilde{a}_\alpha$, we conclude that the expression in the bracket must vanish. We thus have for the $\Gamma^\varrho_{\lambda\mu}$ the following transformation formula:

$$\tilde{\Gamma}^\varrho_{\lambda\mu} = \frac{\partial\tilde{x}^\varrho}{\partial x^\alpha}\frac{\partial x^\beta}{\partial\tilde{x}^\lambda}\frac{\partial x^\gamma}{\partial\tilde{x}^\mu}\Gamma^\alpha_{\beta\gamma} + \frac{\partial\tilde{x}^\varrho}{\partial x^\alpha}\frac{\partial^2 x^\alpha}{\partial\tilde{x}^\lambda\partial\tilde{x}^\mu}. \tag{5.5}$$

One can see at once that (5.5) is not only a necessary, but also a sufficient condition for the expression (5.3) to be a tensor.

Note that the symmetry of $\Gamma^\alpha_{\lambda\mu}$ in λ and μ, which is suggested by the heuristic expression (5.2′), is not necessary: One can verify directly that the expression (5.3) will be a tensor also in the case $\Gamma^{\alpha}_{\lambda\mu} \neq \Gamma^{\alpha}_{\mu\lambda}$, the only condition being that $\Gamma^{\alpha}_{\lambda\mu}$ transforms according to (5.5).

The quantity $\Gamma^{\alpha}_{\lambda\mu}$ is called an (affine) *connection*. It has, in the general case, n^3 independent components. The simplest way to define a connection is by giving its n^3 components in some coordinate system x^{μ} and then calculating the values of the components in any other coordinate system $\tilde{x}^{\mu}$ with the help of (5.5). Note that in (5.5) the first term on the right hand side has the same form as the expression describing the transformation of a tensor of order (1, 2). It is the presence of the last term in (5.5) which has the consequence that $\Gamma^{\alpha}_{\lambda\mu}$ is not a tensor.

The following properties of connections follow immediately from the definition (5.5):

(1) If in a given space there are defined two connections, ${}_1\Gamma^{\alpha}_{\lambda\mu}$ and ${}_2\Gamma^{\alpha}_{\lambda\mu}$, their difference is a tensor:

$$ {}_1\Gamma^{\alpha}_{\lambda\mu} - {}_2\Gamma^{\alpha}_{\lambda\mu} \equiv T^{\alpha}_{\lambda\mu} = \text{a tensor of order } (1, 2). \tag{5.6}$$

This remark shows that giving a second connection is equivalent to giving besides the first connection ${}_1\Gamma^{\alpha}_{\lambda\mu}$ also a tensor $T^{\alpha}_{\lambda\mu}$.

(2) If $\Gamma^{\alpha}_{\lambda\mu}$ is a connection, then $\Gamma^{\alpha}_{\mu\lambda}$ is also a connection.

(3) The expression

$$ \tfrac{1}{2}(\Gamma^{\alpha}_{\lambda\mu} + \Gamma^{\alpha}_{\mu\lambda}) \equiv \Gamma^{\alpha}_{(\lambda\mu)} \tag{5.7}$$

is a symmetric connection.

(4) The expression

$$ \tfrac{1}{2}(\Gamma^{\alpha}_{\lambda\mu} - \Gamma^{\alpha}_{\mu\lambda}) \equiv \Gamma^{\alpha}_{[\lambda\mu]} \tag{5.8}$$

is a tensor which is antisymmetric in λ, μ. This tensor is called the *torsion* of the space.

(5) A general non-symmetric connection $\Gamma^{\alpha}_{\lambda\mu}$ can be written in the form:

$$ \Gamma^{\alpha}_{\lambda\mu} = \Gamma^{\alpha}_{(\lambda\mu)} + \Gamma^{\alpha}_{[\lambda\mu]} \tag{5.9}$$

thus being the sum of a symmetric connection and of a tensor.

The tensor given by Equation (5.3) is called the *absolute derivative* or the *covariant derivative* of the vectorfield a_{λ}. It is customary to denote it by a semicolon:

$$ a_{\lambda;\mu} = a_{\lambda,\mu} - \Gamma^{\alpha}_{\lambda\mu} a_{\alpha}. \tag{5.10}$$

Note that, when the connection is non-symmetric, there is some arbitrariness in the definition (5.10): We could write this relation also with $\Gamma^{\alpha}_{\mu\lambda}$ instead of $\Gamma^{\alpha}_{\lambda\mu}$. For symmetric connections, which interest us in general relativity, this ambiguity does not exist.

A symmetric connection has, in a space of n dimensions, $n^2(n+1)/2$ independent components. The following important theorem is valid for symmetric connections. Let a symmetric connection $\Gamma^{\lambda}_{\mu\nu}$ have in the coordinate system x^{μ} and at the point P the component values $(\Gamma^{\lambda}_{\mu\nu})_P$. It is always possible to find a transformation $x^{\mu} \rightarrow \tilde{x}^{\mu}$ such that

$$ (\tilde{\Gamma}^{\lambda}_{\mu\nu})_P = 0. \tag{5.11}$$

To prove the theorem let us consider a transformation of the form

$$x^\mu - x^\mu_P = \tilde{x}^\mu + \tfrac{1}{2} c^\mu_{\alpha\beta} \tilde{x}^\alpha \tilde{x}^\beta, \tag{5.12}$$

$c^\mu_{\alpha\beta} = c^\mu_{\beta\alpha}$ being constant coefficients and $x^\mu{}_P$ the coordinates of P in the coordinate system x^μ. This transformation is certainly meaningful in some sufficiently small region around P. If we notice that $\tilde{x}^\mu{}_P = 0$, we find at once:

$$\left(\frac{\partial x^\mu}{\partial \tilde{x}^\lambda}\right)_P = \delta^\mu_\lambda = \left(\frac{\partial \tilde{x}^\mu}{\partial x^\lambda}\right)_P ; \qquad \frac{\partial^2 x^\lambda}{\partial \tilde{x}^\mu \partial \tilde{x}^\nu} = c^\lambda_{\mu\nu} .$$

Therefore from (5.5):

$$(\tilde{\Gamma}^\alpha_{\lambda\mu})_P = (\Gamma^\alpha_{\lambda\mu})_P + c^\alpha_{\lambda\mu}, \tag{5.13}$$

and it is sufficient to take $c^\alpha_{\lambda\mu} = -(\Gamma^\alpha_{\lambda\mu})_P$ in order to obtain (5.11).

The following more general result concerning a space with a symmetric connection will be stated here without proof: Given an arbitrary curve in such a space, we can always introduce coordinates in which the $\Gamma^\lambda_{\mu\nu}$ vanish at all points of the curve.

We shall see later that in a Riemannian space there is a symmetric connection, the components of which are the Christoffel symbols, determined directly from the metrical tensor and its derivatives. For the moment we shall consider general, non-symmetric connections.

6. Rules for Covariant Differentiation

The formula for the covariant derivative of a tensor field of any order will be derived from the already known results concerning scalars and covariant vectors:

(1) The covariant derivative of a scalar is, according to Equation (4.3′), identical with its ordinary derivative:

$$\varphi_{;\lambda} = \varphi_{,\lambda} . \tag{6.1}$$

(2) The covariant derivative of a covariant vector is given by Equation (5.10). In order to arrive at uniquely determined expressions we shall demand for the covariant derivative of a product of tensors the same rule as for the ordinary derivative:

$$(A^{\cdots}_{\cdots} B^{\cdots}_{\cdots})_{;\lambda} = A^{\cdots}_{\cdots;\lambda} B^{\cdots}_{\cdots} + A^{\cdots}_{\cdots} B^{\cdots}_{\cdots;\lambda} . \tag{6.2}$$

We consider first a contravariant vectorfield a^λ. If b_λ is some covariant vectorfield, we shall have

$$a^\lambda b_\lambda \equiv \varphi = \text{scalar} .$$

Therefore, according to (6.1) and (6.2):

$$(a^\lambda b_\lambda)_{;\mu} = a^\lambda{}_{;\mu} b_\lambda + a^\lambda b_{\lambda;\mu} = a^\lambda{}_{,\mu} b_\lambda + a^\lambda b_{\lambda,\mu} . \tag{6.2′}$$

Now according to (5.10):

$$a^\lambda b_{\lambda;\mu} = a^\lambda (b_{\lambda,\mu} - \Gamma^\alpha_{\lambda\mu} b_\alpha) .$$

Introducing this result in (6.2′) we find:

$$a^{\lambda}{}_{;\mu}b_{\lambda} = (a^{\lambda}{}_{,\mu} + \Gamma^{\lambda}_{\alpha\mu}a^{\alpha})\, b_{\lambda}\,.$$

Since b_{λ} is arbitrary, we must have:

$$a^{\lambda}{}_{;\mu} = a^{\lambda}{}_{,\mu} + \Gamma^{\lambda}_{\alpha\mu}a^{\alpha}\,. \tag{6.3}$$

Let us now consider a contravariant tensor $T^{\lambda\mu}$ of order 2. With two arbitrary covariant vectors a_{λ} and b_{λ} we obtain a scalar φ:

$$T^{\lambda\mu}a_{\lambda}b_{\mu} = \varphi\,.$$

Differentiating this relation we find:

$$T^{\lambda\mu}{}_{;\nu}a_{\lambda}b_{\mu} + T^{\lambda\mu}(a_{\lambda;\nu}b_{\mu} + a_{\lambda}b_{\mu;\nu}) = T^{\lambda\mu}{}_{,\nu}a_{\lambda}b_{\mu} + T^{\lambda\mu}(a_{\lambda,\nu}b_{\mu} + a_{\lambda}b_{\mu,\nu})\,.$$

Introducing in this relation the values of $a_{\lambda;\nu}$ and $b_{\mu;\nu}$ according to (5.10) we arrive at the result:

$$\{T^{\lambda\mu}{}_{;\nu} - T^{\lambda\mu}{}_{,\nu} - \Gamma^{\lambda}_{\alpha\nu}T^{\alpha\mu} - \Gamma^{\mu}_{\alpha\nu}T^{\lambda\alpha}\}\, a_{\lambda}b_{\mu} = 0\,.$$

Since a_{λ} and b_{μ} are arbitrary vectors, we must have:

$$T^{\lambda\mu}{}_{;\nu} = T^{\lambda\mu}{}_{,\nu} + \Gamma^{\lambda}_{\alpha\nu}T^{\alpha\mu} + \Gamma^{\mu}_{\alpha\nu}T^{\lambda\alpha}\,. \tag{6.4}$$

In the case of a mixed tensor $T^{\lambda}{}_{\mu}$ we shall obtain a scalar using one covariant vector a_{λ} and one contravariant b^{μ}. A simple calculation of the same type as the preceding one, using now the relations (5.10) and (6.3), leads to the following result:

$$T^{\lambda}{}_{\mu;\nu} = T^{\lambda}{}_{\mu,\nu} + \Gamma^{\lambda}_{\alpha\nu}T^{\alpha}{}_{\mu} - \Gamma^{\alpha}_{\mu\nu}T^{\lambda}{}_{\alpha}\,. \tag{6.5}$$

The previous examples allow to write the formula for the covariant derivative of a tensor of any order (p, q):

$$\left.\begin{aligned} T^{\lambda\mu\ldots}{}_{\nu\varrho\ldots;\sigma} = T^{\lambda\mu\ldots}{}_{\nu\varrho\ldots,\sigma} &+ \Gamma^{\lambda}_{\alpha\sigma}T^{\alpha\mu\ldots}{}_{\nu\varrho\ldots} + \Gamma^{\mu}_{\alpha\sigma}T^{\lambda\alpha\ldots}{}_{\nu\varrho\ldots} + \cdots \\ &- \Gamma^{\alpha}_{\nu\sigma}T^{\lambda\mu\ldots}{}_{\alpha\varrho\ldots} - \Gamma^{\alpha}_{\varrho\sigma}T^{\lambda\mu\ldots}{}_{\nu\alpha\ldots} - \cdots\,. \end{aligned}\right\} \tag{6.6}$$

The right-hand side of this equation starts with the ordinary derivative of the tensor. Then we have one term of the type of the last term in (6.3) for each upper index and one term of the type of the last term in (5.10) for each lower index.

As a first application of the preceding formulae let us calculate the covariant derivative of the Kronecker tensorfield $A^{\lambda}{}_{\mu}(x^{\alpha}) = \delta^{\lambda}{}_{\mu}$. We find from (6.5):

$$\delta^{\lambda}{}_{\mu;\nu} = \delta^{\lambda}{}_{\mu,\nu} + \Gamma^{\lambda}_{\alpha\nu}\delta^{\alpha}_{\mu} - \Gamma^{\alpha}_{\mu\nu}\delta^{\lambda}_{\alpha}\,.$$

Since $\delta^{\lambda}{}_{\mu} = \text{const}$ and $\Gamma^{\lambda}_{\alpha\nu}\delta^{\alpha}_{\mu} = \Gamma^{\lambda}_{\mu\nu} = \Gamma^{\alpha}_{\mu\nu}\delta^{\lambda}_{\alpha}$, it follows that

$$\delta^{\lambda}{}_{\mu;\nu} = 0: \tag{6.7}$$

The Kronecker tensor is also covariantly constant.

As a second example we calculate the antisymmetric part of $a_{\lambda;\mu}$. Starting from (5.10) we find at once:

$$a_{[\lambda;\mu]} = a_{[\lambda,\mu]} - \Gamma^{\alpha}_{[\lambda\mu]}a_{\alpha}\,. \tag{6.8}$$

The first and the last term in this equation are tensors. It follows that the term $a_{[\lambda,\mu]}$ will be also a tensor, as we have proved earlier (Section 4) by a direct computation. Note that, if the connection is symmetric, we shall have

$$a_{[\lambda;\mu]} = a_{[\lambda,\mu]}. \tag{6.8'}$$

We shall now discuss the differentiation of tensor-densities. In this case the basic question is the following: How can we construct a covariant vector density by differentiating a scalar density Q? We shall consider densities of weight $w=1$ only.

From the transformation formula (3.3),

$$\tilde{Q} = QD, \qquad D = \det \frac{\partial x^\alpha}{\partial \tilde{x}^\lambda},$$

we derive at once the relation:

$$\frac{\partial \tilde{Q}}{\partial \tilde{x}^\mu} = \frac{\partial Q}{\partial x^\beta}\frac{\partial x^\beta}{\partial \tilde{x}^\mu} D + Q \frac{\partial D}{\partial \tilde{x}^\mu}. \tag{6.9}$$

In order to determine the derivative of D we have to use the following formula for the variation of a determinant:

$$\delta(\det a^\alpha{}_\lambda) = \delta a^\alpha{}_\lambda \cdot A^\lambda{}_\alpha. \tag{6.10}$$

$A^\lambda{}_\alpha$ is the minor corresponding to the element $a^\alpha{}_\lambda$. Therefore we shall have the relation:

$$a^\alpha{}_\lambda A^\lambda{}_\beta = \delta^\alpha{}_\beta \cdot \det a^\lambda{}_\mu. \tag{6.11}$$

Equation (6.10) is equivalent to:

$$\frac{\partial}{\partial \tilde{x}^\mu}(\det a^\alpha{}_\lambda) = \frac{\partial}{\partial \tilde{x}^\mu} a^\alpha{}_\lambda \cdot A^\lambda{}_\alpha. \tag{6.10'}$$

In the case of the determinant D we have

$$a^\alpha{}_\lambda = \frac{\partial x^\alpha}{\partial \tilde{x}^\lambda}. \tag{6.12}$$

Comparing (6.11) and (6.12) with (1.6) we see that in this case it is:

$$A^\lambda{}_\beta = \frac{\partial \tilde{x}^\lambda}{\partial x^\beta} \cdot D. \tag{6.13}$$

Introducing (6.12) and (6.13) in (6.10') we find:

$$\frac{\partial D}{\partial \tilde{x}^\mu} = \frac{\partial^2 x^\alpha}{\partial \tilde{x}^\lambda \partial \tilde{x}^\mu}\frac{\partial \tilde{x}^\lambda}{\partial x^\alpha} \cdot D. \tag{6.14}$$

Consider now the transformation formula (5.5) for the connection and apply to it the contraction $\varrho=\lambda$. The result is:

$$\tilde{\Gamma}^\varrho_{\varrho\mu} = \frac{\partial x^\gamma}{\partial \tilde{x}^\mu}\Gamma^\alpha_{\alpha\gamma} + \frac{\partial^2 x^\alpha}{\partial \tilde{x}^\lambda \partial \tilde{x}^\mu}\frac{\partial \tilde{x}^\lambda}{\partial x^\alpha}. \tag{6.15}$$

With this result we rewrite Equation (6.14) in the form:

$$\frac{\partial D}{\partial \tilde{x}^{\mu}} = \left(\tilde{\Gamma}^{\varrho}_{\varrho\mu} - \frac{\partial x^{\alpha}}{\partial \tilde{x}^{\mu}} \Gamma^{\varrho}_{\varrho\alpha} \right) D . \tag{6.14'}$$

Introducing this expression in (6.9) we find:

$$\frac{\partial \tilde{Q}}{\partial \tilde{x}^{\mu}} = \frac{\partial Q}{\partial x^{\alpha}} \frac{\partial x^{\alpha}}{\partial \tilde{x}^{\mu}} \cdot D + Q \left(\tilde{\Gamma}^{\alpha}_{\alpha\mu} - \frac{\partial x^{\beta}}{\partial \tilde{x}^{\mu}} \Gamma^{\alpha}_{\alpha\beta} \right) D .$$

Or finally:

$$\left(\frac{\partial \tilde{Q}}{\partial \tilde{x}^{\mu}} - \tilde{Q} \tilde{\Gamma}^{\alpha}_{\alpha\mu} \right) = \frac{\partial x^{\beta}}{\partial \tilde{x}^{\mu}} \left(\frac{\partial Q}{\partial x^{\beta}} - Q \Gamma^{\alpha}_{\alpha\beta} \right) D . \tag{6.16}$$

This means, according to (3.6), that the quantity $(\partial Q/\partial x^{\beta}) - Q\Gamma^{\alpha}_{\alpha\beta}$ is a covariant vector-density. We shall call it the covariant derivative of the scalar density Q:

$$Q_{;\beta} = Q_{,\beta} - \Gamma^{\alpha}_{\alpha\beta} Q . \tag{6.17}$$

Remembering that any tensor density is the product of a tensor by a scalar density we can now derive formulae for the covariant derivatives of tensor densities of arbitrary order and of weight $w=1$. It will be sufficient to extend to tensor densities the rule (6.2) and then to use the results for the covariant derivatives of tensors combined with (6.17). We shall give two examples.

Let $\mathfrak{A}_{\mu}$ be a covariant vector-density. If b^{μ} is an arbitrary contravariant vector, we shall have:

$$\mathfrak{A}_{\mu} b^{\mu} \equiv Q = \text{scalar density} .$$

Now from (6.2) and (6.17):

$$(\mathfrak{A}_{\mu} b^{\mu})_{;\nu} = \mathfrak{A}_{\mu;\nu} b^{\mu} + \mathfrak{A}_{\mu} b^{\mu}{}_{;\nu} = (\mathfrak{A}_{\mu} b^{\mu})_{,\nu} - \Gamma^{\alpha}_{\alpha\nu} \mathfrak{A}_{\mu} b^{\mu} .$$

Introducing in this relation the expression (6.3) for $b^{\mu}{}_{;\nu}$ we find finally:

$$\mathfrak{A}_{\mu;\nu} b^{\mu} = (\mathfrak{A}_{\mu,\nu} - \Gamma^{\alpha}_{\mu\nu} \mathfrak{A}_{\alpha} - \Gamma^{\alpha}_{\alpha\nu} \mathfrak{A}_{\mu}) \, b^{\mu} .$$

Since b^{μ} is arbitrary, we shall have:

$$\mathfrak{A}_{\mu;\nu} = \mathfrak{A}_{\mu,\nu} - \Gamma^{\alpha}_{\mu\nu} \mathfrak{A}_{\alpha} - \Gamma^{\alpha}_{\alpha\nu} \mathfrak{A}_{\mu} . \tag{6.18}$$

This is the formula for the covariant derivative of the vector-density $\mathfrak{A}_{\mu}$.

In the case of a contravariant vector-density $\mathfrak{A}^{\mu}$ we have to use an arbitrary covariant vector b_{μ}. Proceeding in the same way we find:

$$\mathfrak{A}^{\mu}{}_{;\nu} = \mathfrak{A}^{\mu}{}_{,\nu} + \Gamma^{\mu}_{\alpha\nu} \mathfrak{A}^{\alpha} - \Gamma^{\alpha}_{\alpha\nu} \mathfrak{A}^{\mu} . \tag{6.19}$$

We write also the formula for the derivative of a tensor density $\mathfrak{A}^{\lambda\cdot\cdot}{}_{\mu\cdot\cdot}$ of any order:

$$\mathfrak{A}^{\lambda\cdot\cdot}{}_{\mu\cdot\cdot;\varrho} = \mathfrak{A}^{\lambda\cdot\cdot}{}_{\mu\cdot\cdot,\varrho} + \Gamma^{\lambda}_{\alpha\varrho} \mathfrak{A}^{\alpha\cdot\cdot}{}_{\mu\cdot\cdot} + \cdots - \Gamma^{\alpha}_{\mu\varrho} \mathfrak{A}^{\lambda\cdot\cdot}{}_{\alpha\cdot\cdot} - \cdots - \Gamma^{\alpha}_{\alpha\varrho} \mathfrak{A}^{\lambda\cdot\cdot}{}_{\mu\cdot\cdot} . \tag{6.20}$$

Comparing this formula with (6.6) we see that they are identical in form up to the last term in (6.20).

As an application of Equation (6.20) we calculate the derivative of the field $\mathfrak{A}^{\lambda\mu\nu\varrho}(x^\alpha)=\varepsilon^{\lambda\mu\nu\varrho}$, $\varepsilon^{\lambda\mu\nu\varrho}$ being the Levi-Civita tensor density defined in Section 3 (space of $n=4$ dimensions). We have:

$$\varepsilon^{\lambda\mu\nu\varrho}{}_{;\sigma} = \varepsilon^{\lambda\mu\nu\varrho}{}_{,\sigma} + \Gamma^{\lambda}_{\alpha\sigma}\varepsilon^{\alpha\mu\nu\varrho} + \Gamma^{\mu}_{\alpha\sigma}\varepsilon^{\lambda\alpha\nu\varrho} + \Gamma^{\nu}_{\alpha\sigma}\varepsilon^{\lambda\mu\alpha\varrho} + \Gamma^{\varrho}_{\alpha\sigma}\varepsilon^{\lambda\mu\nu\alpha} - \Gamma^{\alpha}_{\alpha\sigma}\varepsilon^{\lambda\mu\nu\varrho}. \tag{6.21}$$

The first term on the right-hand side vanishes because $\varepsilon^{\lambda\mu\nu\varrho}=\text{const}$. To calculate the sum of the remaining terms we recall that this expression is antisymmetric in λ, μ, ν, ϱ and consequently it vanishes if the values of λ, μ, ν, ϱ are not all different. It is therefore sufficient to consider the component of Equation (6.21) corresponding to $\lambda=1$, $\mu=2$, $\nu=3$, $\varrho=4$. In this case we find at once:

$$\Gamma^{\lambda}_{\alpha\sigma}\varepsilon^{\alpha\mu\nu\varrho} = \Gamma^{1}_{\alpha\sigma}\varepsilon^{\alpha 234} = \Gamma^{1}_{1\sigma}\varepsilon^{1234} = \Gamma^{1}_{1\sigma};$$

$$\Gamma^{\mu}_{\alpha\sigma}\varepsilon^{\lambda\alpha\nu\varrho} = \Gamma^{2}_{2\sigma}, \qquad \Gamma^{\nu}_{\alpha\sigma}\varepsilon^{\lambda\mu\alpha\varrho} = \Gamma^{3}_{3\sigma}, \qquad \Gamma^{\varrho}_{\alpha\sigma}\varepsilon^{\lambda\mu\nu\alpha} = \Gamma^{4}_{4\sigma}.$$

Therefore the right-hand side of Equation (6.21) is

$$\Gamma^{1}_{1\sigma} + \Gamma^{2}_{2\sigma} + \Gamma^{3}_{3\sigma} + \Gamma^{4}_{4\sigma} - \Gamma^{\alpha}_{\alpha\sigma} = 0.$$

Consequently the covariant derivative of $\varepsilon^{\lambda\mu\nu\varrho}$ vanishes:

$$\varepsilon^{\lambda\mu\nu\varrho}{}_{;\sigma} = 0. \tag{6.22}$$

7. Parallel Transport

Let a_λ be some covariant vector field. Consider two neighbouring points P and P', the displacement **PP**$'$ having the components dx^μ. We shall then have:

$$a_\mu(P') - a_\mu(P) = a_{\mu,\nu}\,dx^\nu. \tag{7.1}$$

This relation has the following meaning. The derivative $a_{\mu,\nu}$ is the coefficient of dx^ν in the difference $a_\mu(P')-a_\mu(P)$. In Section 2 we stressed that the sum (or the difference) of two tensors is again a tensor on condition that the two tensors be given at the same point. This is not the case for the difference $a_\mu(P')-a_\mu(P)$ and this suggests why the ordinary derivative $a_{\lambda,\mu}$ is not a tensor.

Now the connection $\Gamma^{\lambda}_{\mu\nu}$ allows us to define the covariant derivative $a_{\lambda;\mu}$ which is a tensor. The meaning of this fact is that with the help of $\Gamma^{\lambda}_{\mu\nu}$ we can determine a vector at the point P', which has to be considered as equivalent to the vector a_λ given at P. In other words the connection $\Gamma^{\lambda}_{\mu\nu}$ allows to define the 'transport' of the vector a_λ from a point P to the neighbouring point P'. For reasons which will be explained later we call this operation the *parallel transport* defined by the connection $\Gamma^{\lambda}_{\mu\nu}$.

The formula giving the quantitative description of this operation will be derived as follows. Let a_λ be the components of the vector given at P and $a_\lambda+\delta a_\lambda$ the components of this vector after its parallel transport to P'. A second vector is defined at P' by the vector field $a_\lambda(x^\alpha)$. The components of this second vector are:

$$a_\lambda + da_\lambda = a_\lambda + a_{\lambda,\mu}\,dx^\mu.$$

The difference of the two vectors defined at P' will be a vector, which is related to the tensor $a_{\lambda;\mu}$ as follows:

$$(a_\lambda + \mathrm{d}a_\lambda) - (a_\lambda + \delta a_\lambda) \equiv \mathrm{d}a_\lambda - \delta a_\lambda = a_{\lambda;\mu}\,\mathrm{d}x^\mu\,. \tag{7.2}$$

Introducing in this relation the expression (5.10) for $a_{\lambda;\mu}$ we find at once the quantity δa_λ:

$$\delta a_\lambda = \Gamma^\nu_{\lambda\mu} a_\nu\,\mathrm{d}x^\mu\,. \tag{7.3}$$

Note that this expression is linear in a_ν and $\mathrm{d}x^\mu$ as well as in $\Gamma^\nu_{\lambda\mu}$. Note also that, while δa_λ and $\mathrm{d}a_\lambda$ are not vectors separately, the difference $\mathrm{d}a_\lambda - \delta a_\lambda$ is a vector according to (7.2).

In a similar way we can discuss the parallel transport of a contravariant vector. Let a^λ be the components of the vector given at P and $a^\lambda + \delta a^\lambda$ the components of the vector after its transport to P'. At P' we have also the vector belonging to the field $a^\lambda(x^\alpha)$, the components of which are

$$a^\lambda(P') = a^\lambda + \mathrm{d}a^\lambda = a^\lambda + a^\lambda{}_{,\mu}\,\mathrm{d}x^\mu\,.$$

The difference of the two vectors given at P' will be related to the absolute derivative $a^\lambda{}_{;\mu}$:

$$(a^\lambda + \mathrm{d}a^\lambda) - (a^\lambda + \delta a^\lambda) = \mathrm{d}a^\lambda - \delta a^\lambda = a^\lambda{}_{;\mu}\,\mathrm{d}x^\mu\,. \tag{7.4}$$

Introducing in this relation the expression (6.3) for $a^\lambda{}_{;\mu}$ we find:

$$\delta a^\lambda = -\,\Gamma^\lambda_{\mu\nu} a^\mu\,\mathrm{d}x^\nu\,. \tag{7.5}$$

The formula for the parallel transport of any other tensor can be derived in the same way. Especially interesting as well as simple is the case of a scalar φ. In this case we find at once

$$\delta\varphi = 0, \tag{7.6}$$

this being a consequence of Equation (6.1).

An interesting scalar is the scalar product of a contravariant vector a^λ by a covariant one b_λ:

$$\varphi = a^\lambda b_\lambda\,.$$

In this case the result (7.6) follows also from the Equations (7.3) and (7.5). Indeed, we find in this case:

$$\delta(a^\lambda b_\lambda) = \delta a^\lambda \cdot b_\lambda + a^\lambda \delta b_\lambda = -\,\Gamma^\lambda_{\mu\nu} a^\mu b_\lambda\,\mathrm{d}x^\nu + a^\lambda \Gamma^\mu_{\lambda\nu} b_\mu\,\mathrm{d}x^\nu = 0\,. \tag{7.7}$$

The relation (7.7) generalises an elementary result of Euclidean geometry: In Euclidean space the parallel transport of two vectors leaves their scalar product unchanged. Euclidean space is a special example of a Riemannian space. Parallel transport in a Riemannian space will be discussed in some detail later. Here we note that the result (7.7) remains valid more generally in any connected space.

The operation of parallel transport can be applied to a vector also along a curve leading from the point P to some other point Q. The change of the components of the vector can be calculated in this case by integration of the expression (7.3) or (7.5) along the curve. The following important remark should be made here. The result of parallel transport from P to Q will in general depend not only on the endpoints P and Q but also on the curve connecting the two points. The result will be independent of the curve only in one special case. This special case is characterised by a condition which will be derived in Section 9.

8. Geodesics

In a space of n dimensions a curve is determined by n equations of the form

$$x^\mu = f^\mu(\lambda)\,; \qquad \mu = 1, 2, \ldots, n\,, \tag{8.1}$$

λ being some parameter which we shall assume to be a scalar. At a point P of the curve let a^λ be a contravariant vector. If we apply to this vector the operation of parallel transport along the curve, we shall obtain a vector A^λ at each point Q of the curve: A^λ is the result of the parallel transport of a^λ from P to Q along the curve. (We shall have $A^\lambda = a^\lambda$ at P.)

We now assume that a^λ is the tangent vector of the curve at P:

$$a^\mu = \frac{\mathrm{d}x^\mu}{\mathrm{d}\lambda} \quad \text{at} \quad P\,. \tag{8.2}$$

Parallel transport of this vector will determine at a point Q of the curve a vector A^μ, which in general will not be tangent to the curve. There is, however, a special case in which we shall have the following situation: When a^μ is tangent to the curve at P, the vector A^μ is tangent to the curve at Q, for any point Q of the curve. In this case we call the curve a geodesic curve or simply a *geodesic* of the space.

We shall now derive the condition which characterises a geodesic. Consider two neighbouring points P and P' on the curve, corresponding to the values λ and $\lambda + \mathrm{d}\lambda$ of the parameter. At P we have the tangent vector (8.2). The tangent vector at P' has the components

$$a^\mu(P') = \frac{\mathrm{d}x^\mu}{\mathrm{d}\lambda} + \frac{\mathrm{d}^2x^\mu}{\mathrm{d}\lambda^2}\,\mathrm{d}\lambda\,. \tag{8.3}$$

If we apply the operation of parallel transport to the vector $a^\mu(P)$ from P to P', we shall have at P' a second vector, the components of which will be according to (7.5):

$$\frac{\mathrm{d}x^\mu}{\mathrm{d}\lambda} + \delta\left(\frac{\mathrm{d}x^\mu}{\mathrm{d}\lambda}\right) = \frac{\mathrm{d}x^\mu}{\mathrm{d}\lambda} - \Gamma^\mu_{\varrho\sigma}\frac{\mathrm{d}x^\varrho}{\mathrm{d}\lambda}\frac{\mathrm{d}x^\sigma}{\mathrm{d}\lambda}\,\mathrm{d}\lambda\,. \tag{8.4}$$

The curve will be a geodesic if the two vectors (8.3) and (8.4) are collinear, i.e. if the components (8.3) are proportional to (8.4). The proportionality factor will evidently differ from 1 by a quantity of the first order in $\mathrm{d}\lambda$. Writing it in the form $1 + f(\lambda)\,\mathrm{d}\lambda$ we have:

$$\frac{\mathrm{d}x^\mu}{\mathrm{d}\lambda} + \frac{\mathrm{d}^2x^\mu}{\mathrm{d}\lambda^2}\,\mathrm{d}\lambda = \{1 + f(\lambda)\,\mathrm{d}\lambda\}\left(\frac{\mathrm{d}x^\mu}{\mathrm{d}\lambda} - \Gamma^\mu_{\varrho\sigma}\frac{\mathrm{d}x^\varrho}{\mathrm{d}\lambda}\frac{\mathrm{d}x^\sigma}{\mathrm{d}\lambda}\,\mathrm{d}\lambda\right).$$

This leads to the equation

$$\frac{d^2x^\mu}{d\lambda^2} + \Gamma^\mu_{\varrho\sigma}\frac{dx^\varrho}{d\lambda}\frac{dx^\sigma}{d\lambda} = f(\lambda)\frac{dx^\mu}{d\lambda}, \tag{8.5}$$

which must of course be satisfied at every point of the curve. Equation (8.5) is the general form of the differential equation of a geodesic.

The following two remarks can be made about Equation (8.5):

(1) A solution of this equation, i.e. a geodesic, will be completely determined by a point P and the direction of the tangent vector at P.

(2) If the connection $\Gamma^\lambda_{\varrho\sigma}$ is non-symmetric, then according to (5.9):

$$\Gamma^\lambda_{\varrho\sigma} = \Gamma^\lambda_{(\varrho\sigma)} + \Gamma^\lambda_{[\varrho\sigma]}.$$

One verifies at once that only the symmetric part $\Gamma^\lambda_{(\varrho\sigma)}$ contributes to Equation (8.5): The antisymmetric part will cancel because the factor $(dx^\varrho/d\lambda)\,(dx^\sigma/d\lambda)$ is symmetric in ϱ, σ.

We can obtain a simpler form of the Equation (8.5) by the use of a special class of parameters σ. In order to prove this let us introduce in (8.5) a new parameter σ by an equation of the form

$$\lambda = \lambda(\sigma). \tag{8.6}$$

We shall then have:

$$\frac{dx^\mu}{d\lambda} = \frac{dx^\mu}{d\sigma}\frac{d\sigma}{d\lambda}, \qquad \frac{d^2x^\mu}{d\lambda^2} = \frac{d^2x^\mu}{d\sigma^2}\left(\frac{d\sigma}{d\lambda}\right)^2 + \frac{dx^\mu}{d\sigma}\frac{d^2\sigma}{d\lambda^2}.$$

The Equation (8.5) takes now the following form:

$$\left\{\frac{d^2x^\mu}{d\sigma^2} + \Gamma^\mu_{\alpha\beta}\frac{dx^\alpha}{d\sigma}\frac{dx^\beta}{d\sigma}\right\}\left(\frac{d\sigma}{d\lambda}\right)^2 + \frac{dx^\mu}{d\sigma}\frac{d^2\sigma}{d\lambda^2} = f(\lambda)\frac{dx^\mu}{d\sigma}\frac{d\sigma}{d\lambda}.$$

Therefore if we choose the new parameter σ so as to satisfy the condition

$$\frac{d^2\sigma}{d\lambda^2} = f(\lambda)\frac{d\sigma}{d\lambda}, \tag{8.7}$$

the equation of the geodesic reduces to

$$\frac{d^2x^\mu}{d\sigma^2} + \Gamma^\mu_{\alpha\beta}\frac{dx^\alpha}{d\sigma}\frac{dx^\beta}{d\sigma} = 0. \tag{8.8}$$

A parameter σ of this kind is called an *affine parameter* of the geodesic. It plays in these spaces the role of the proper length in the Minkowski or in any riemannian space.

If λ is already an affine parameter, $\lambda = \tilde{\sigma}$, then the condition (8.7) reduces to

$$\frac{d^2\sigma}{d\tilde{\sigma}^2} = 0.$$

Therefore any other affine parameter σ will be related to $\tilde{\sigma}$ by the simple formula

$$\sigma = a\tilde{\sigma} + b, \quad a \text{ and } b \text{ constants.} \tag{8.9}$$

9. The Curvature Tensor

From the connection $\Gamma^{\lambda}_{\mu\nu}$ we can construct an important tensor, the *curvature tensor* of the space. The actual definition is extremely simple. We start with some covariant vector field a_λ and we compute the second covariant derivative $a_{\lambda;\mu;\nu}$. Remembering that $a_{\lambda;\mu}$ is a tensor, we find from (6.6):

$$a_{\lambda;\mu;\nu} = a_{\lambda;\mu,\nu} - \Gamma^{\varrho}_{\lambda\nu} a_{\varrho;\mu} - \Gamma^{\varrho}_{\mu\nu} a_{\lambda;\varrho}.$$

Introducing in this relation the expressions for the first covariant derivative from (5.10) we find:

$$a_{\lambda;\mu;\nu} = a_{\lambda,\mu,\nu} - \Gamma^{\varrho}_{\lambda\mu,\nu} a_{\varrho} - \Gamma^{\varrho}_{\lambda\mu} a_{\varrho,\nu} - \Gamma^{\varrho}_{\lambda\nu} a_{\varrho,\mu} + \Gamma^{\varrho}_{\lambda\nu}\Gamma^{\sigma}_{\varrho\mu} a_{\sigma} - \Gamma^{\varrho}_{\mu\nu} a_{\lambda;\varrho}. \tag{9.1}$$

We now take the part of this relation which is antisymmetric in μ, ν. The result is:

$$a_{\lambda;\mu;\nu} - a_{\lambda;\nu;\mu} = R^{\varrho}{}_{\lambda\mu\nu} a_{\varrho} - 2\Gamma^{\varrho}_{[\mu\nu]} a_{\lambda;\varrho}; \tag{9.2}$$

$$R^{\varrho}{}_{\lambda\mu\nu} = -\Gamma^{\varrho}_{\lambda\mu,\nu} + \Gamma^{\varrho}_{\lambda\nu,\mu} - \Gamma^{\sigma}_{\lambda\mu}\Gamma^{\varrho}_{\sigma\nu} + \Gamma^{\sigma}_{\lambda\nu}\Gamma^{\varrho}_{\sigma\mu}. \tag{9.3}$$

In Equation (9.2) the left-hand side as well as the last term are tensors. Therefore the term $R^{\varrho}{}_{\lambda\mu\nu} a_{\varrho}$ is also a tensor and, since a_{ϱ} is an arbitrary vector, the quantity $R^{\varrho}{}_{\lambda\mu\nu}$ will be a tensor: This is the curvature tensor of the space. Note that $R^{\varrho}{}_{\lambda\mu\nu}$ depends on nothing else but the connection. It is of order (1, 3) and antisymmetric in the last two indices:

$$R^{\varrho}{}_{\lambda\mu\nu} = -R^{\varrho}{}_{\lambda\nu\mu}. \tag{9.4}$$

We can also start with a contravariant vector b^λ. Proceeding as before we arrive then at the following useful relation:

$$b^{\varrho}{}_{;\mu;\nu} - b^{\varrho}{}_{;\nu;\mu} = -R^{\varrho}{}_{\lambda\mu\nu} b^{\lambda} - 2\Gamma^{\lambda}_{[\mu\nu]} b^{\varrho}{}_{;\lambda} \tag{9.5}$$

with the tensor $R^{\varrho}{}_{\lambda\mu\nu}$ defined again by Equation (9.3).

An important formula containing the curvature tensor will be obtained if we consider the following problem. Let dx^μ and δx^μ be two infinitesimal displacements at the point P (Figure 1) and consider the 'parallelogram' $PABC$. Let a^λ be some vector at P and apply to it the operation of parallel transport along the closed path $PABCP$. The result will be a new vector at P having the components $a^\lambda + \delta a^\lambda$. What are the values of δa^λ?

The quantities δa^λ will be determined by integration of the expression (7.5) along the closed path $PABCP$. A simple calculation, which however we shall not give here, leads to the following result: δa^λ is a quantity of second order given by the formula:

$$\delta a^\lambda = -\tfrac{1}{2} R^{\lambda}{}_{\varrho\mu\nu} a^{\varrho} (dx^\mu \delta x^\nu - dx^\nu \delta x^\mu), \tag{9.6}$$

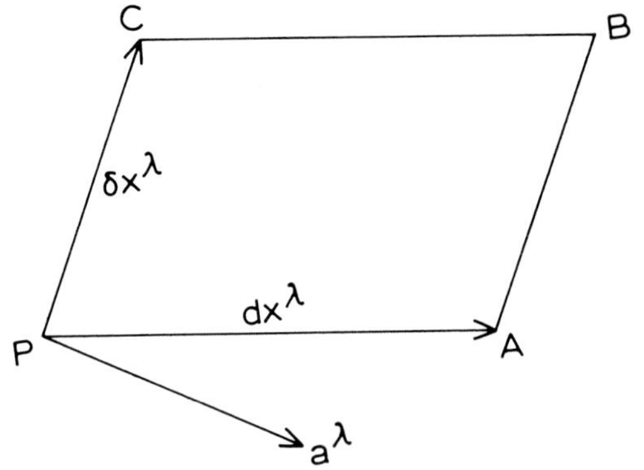

Fig. 1. Parallel transport along a closed path.

$R^{\lambda}{}_{\varrho\mu\nu}$ being the curvature tensor at P.

The formula (9.6) can be integrated over arbitrary 2-dimensional surfaces and it allows then to prove the following important theorem. Consider a region V of the space and in it pairs of points P, Q connected by paths lying entirely in V. Then the necessary and sufficient condition in order that the result of the parallel transport from P to Q be path-independent is that the curvature vanish at all points of V:

$$R^{\varrho}{}_{\lambda\mu\nu} = 0 \quad \text{in} \quad V. \tag{9.7}$$

If the connection $\Gamma^{\lambda}_{\mu\nu}$ is symmetric, $\Gamma^{\lambda}_{\mu\nu} = \Gamma^{\lambda}_{(\mu\nu)}$, one verifies immediately that the curvature tensor defined by Equation (9.3) satisfies the following additional relation:

$$R^{\varrho}{}_{[\lambda\mu\nu]} = 0. \tag{9.8}$$

The following theorem is valid in this case. If the connection is symmetric and the corresponding tensor $R^{\varrho}{}_{\lambda\mu\nu}$ vanishes in a region V, one can obtain

$$\tilde{\Gamma}^{\lambda}_{\mu\nu} = 0 \quad \text{in} \quad V \tag{9.9}$$

through an appropriate coordinate transformation $x^{\mu} \to \tilde{x}^{\mu}$. We say in this case that the space is *flat* in the region V.

The curvature tensor being of order (1, 3) ,we can construct contractions of it. They are the following:

$$R^{\lambda}{}_{\lambda\mu\nu} \equiv A_{\mu\nu} = - A_{\nu\mu}\,; \tag{9.10}$$

$$R^{\lambda}{}_{\mu\nu\lambda} \equiv R_{\mu\nu}. \tag{9.11}$$

The contraction $R^{\lambda}{}_{\mu\lambda\nu}$ does not give anything new, because of the relation (9.4):

$$R^{\lambda}{}_{\mu\lambda\nu} = - R^{\lambda}{}_{\mu\nu\lambda} = - R_{\mu\nu}.$$

No other contraction exists in a space which has no metric.

Exercises

II1: Given a tensor $a_{\lambda\mu} = a_{[\lambda\mu]}$ prove that $a_{[\lambda\mu,\nu]}$ is a tensor by using the formula for $a_{[\lambda\mu;\nu]}$. Similar result for $b_{\lambda\mu\nu} = b_{[\lambda u\nu]}$.

II2: Derive the formula for the covariant derivative of a scalar density of weight w (w being a positive or negative integer).

II3: Prove that in a 4-dimensional space the antisymmetric tensor density $\varepsilon_{\lambda\mu\nu\varrho}$ of weight -1 having $\varepsilon_{1234} = 1$ satisfies the relation $\varepsilon_{\lambda\mu\nu\varrho;\sigma} = 0$.

II4: Verify the result (9.6) by direct computation using (7.5).

II5: Obtain a formula analogous to (9.6) for a covariant vector a_λ.

CHAPTER III

RIEMANNIAN GEOMETRY

10. Riemannian Space

A space is called a *metric space* if a prescription is given attributing a scalar *distance* to each pair of neighbouring points. We mention two familiar examples.

(1) The Euclidean space with $n=3$. When we use Cartesian coordinates X^α ($\alpha = 1, 2, 3$), the distance $d\sigma$ of the points $P(X^\alpha)$ and $P'(X^\alpha + dX^\alpha)$ is given by

$$d\sigma^2 = (dX^1)^2 + (dX^2)^2 + (dX^3)^2 . \tag{10.1}$$

(2) The Minkowski space of special relativity. In an inertial coordinate system we have

$$ds^2 = (dX^1)^2 + (dX^2)^2 + (dX^3)^2 - (dX^4)^2 . \tag{10.2}$$

In both these cases the expression, which determines the distance, is a sum of the squares of the differentials dX^α multiplied by ± 1.

If we introduce general coordinates x^μ by the equations

$$X^\mu = f^\mu(x^\alpha),$$

we shall have:

$$dX^\mu = \frac{\partial X^\mu}{\partial x^\alpha} dx^\alpha .$$

With these new coordinates we shall have for $d\sigma^2$ or ds^2 an expression, which is homogeneous and quadratic in the dx^α. I.e. the general expression for ds^2 will be of the following form:

$$ds^2 = g_{\mu\nu}\, dx^\mu\, dx^\nu . \tag{10.3}$$

In order to have the $g_{\mu\nu}$ uniquely determined we must require that $g_{\mu\nu}$ be symmetric:

$$g_{\mu\nu} = g_{\nu\mu} = g_{(\mu\nu)} . \tag{10.4}$$

Indeed, if we were to allow an antisymmetric part $g_{[\mu\nu]} \neq 0$, this part would play no role in (10.3) because of the symmetry in μ, ν of the product $dx^\mu\, dx^\nu$.

The relation (10.3) characterises a *Riemannian space*: This is a metric space in which the distance between neighbouring points is given by a formula of the type (10.3). Thus Euclidean as well as Minkowski space are special cases of Riemannian space. In a general riemannian space the $g_{\mu\nu}$ can be arbitrarily given functions of the coordinates and it is not possible to reduce them by a coordinate transformation to a simple form of the kind appearing in (10.1) or (10.2). Such a reduction is possible only if the $g_{\mu\nu}$ satisfy a condition which we shall derive in Section 14.

The functions $g_{\mu\nu}(x^\alpha)$ are the components of a covariant tensorfield of order 2. This follows immediately from the fact that $\mathrm{d}s^2$ is a scalar. Indeed, if we consider a second coordinate system $\tilde{x}^\mu$ defined by the equations

$$x^\mu = f^\mu(\tilde{x}^\alpha),$$

we shall have

$$\mathrm{d}x^\mu = \frac{\partial x^\mu}{\partial \tilde{x}^\alpha}\,\mathrm{d}\tilde{x}^\alpha;$$

$$\mathrm{d}s^2 = g_{\mu\nu}\,\mathrm{d}x^\mu\,\mathrm{d}x^\nu = g_{\mu\nu}\frac{\partial x^\mu}{\partial \tilde{x}^\alpha}\frac{\partial x^\nu}{\partial \tilde{x}^\beta}\,\mathrm{d}\tilde{x}^\alpha\,\mathrm{d}\tilde{x}^\beta \equiv \tilde{g}_{\alpha\beta}\,\mathrm{d}\tilde{x}^\alpha\,\mathrm{d}\tilde{x}^\beta$$

with

$$\tilde{g}_{\alpha\beta} = \frac{\partial x^\mu}{\partial \tilde{x}^\alpha}\frac{\partial x^\nu}{\partial \tilde{x}^\beta}g_{\mu\nu}. \tag{10.5}$$

Equation (10.5) is identical with the transformation formula (1.15) for a covariant tensor of order 2.

The tensor $g_{\mu\nu}$ is called the *metric* tensor or simply the *metric* of the riemannian space. Because of the symmetry (10.4) the tensor $g_{\mu\nu}$ has $n(n+1)/2$ independent components. The metrical properties of the space will be determined completely if we fix in some coordinate system x^α the $n(n+1)/2$ independent functions $g_{\mu\nu}(x^\alpha)$.

If at some point P we are given two infinitesimal displacements ${}_1\mathrm{d}x^\mu$ and ${}_2\mathrm{d}x^\mu$, the metric tensor allows to construct the scalar $g_{\mu\nu}\,{}_1\mathrm{d}x^\mu\,{}_2\mathrm{d}x^\nu$. Generalising a definition from Euclidean space we shall call this scalar the *scalar product* of the two vectors. In the general case of a non-metrical space we could form the scalar product only if one vector were contravariant and the other covariant. The fact that now we can form the product of two contravariant vectors means that in a Riemannian space the metrical tensor allows us to define a covariant vector equivalent to a contravariant one, or in other words to *lower an upper index*:

$$g_{\mu\nu}v^\nu = v_\mu. \tag{10.6}$$

The same operation can be applied to tensors:

$$g_{\mu\alpha}T^{\lambda\alpha} = T^\lambda{}_\mu; \qquad g_{\lambda\alpha}g_{\mu\beta}T^{\alpha\beta} = T_{\lambda\mu} \quad \text{etc.} \tag{10.7}$$

Thus the fundamental distinction between contravariant and covariant tensors does not exist in Riemannian spaces.

In analogy with what is valid in Euclidean space we postulate that the scalar product of two vectors defines the *angle* θ which is formed by them:

$$\mathrm{d}s_{12}^2 = g_{\mu\nu}\,{}_1\mathrm{d}x^\mu\,{}_2\mathrm{d}x^\nu = \mathrm{d}s_1\,\mathrm{d}s_2\cos\theta; \tag{10.8}$$

$$\mathrm{d}s_1^2 = g_{\mu\nu}\,{}_1\mathrm{d}x^\mu\,{}_1\mathrm{d}x^\nu, \qquad \mathrm{d}s_2^2 = g_{\mu\nu}\,{}_2\mathrm{d}x^\mu\,{}_2\mathrm{d}x^\nu. \tag{10.8'}$$

Note, however, that in spaces with indefinite metric, as are Minkowski space and the Riemannian spaces used in general relativity, the angle θ is defined only for vectors which are both non-null.

To understand the meaning of the non-diagonal $g_{\mu\nu}$, e.g. of g_{12}, let us consider the two displacements (in a space with $n=3$)

$$_1\mathrm{d}x^\mu = (\mathrm{d}x^1, 0, 0), \qquad _2\mathrm{d}x^\mu = (0, \mathrm{d}x^2, 0).$$

Their scalar product is, according to (10.8):

$$g_{\mu\nu}\, {}_1\mathrm{d}x^\mu\, {}_2\mathrm{d}x^\nu = g_{12}\, \mathrm{d}x^1\, \mathrm{d}x^2 = \mathrm{d}s_1\, \mathrm{d}s_2 \cos\theta .$$

Therefore we shall have $g_{12} \neq 0$ if $\cos\theta \neq 0$. The vector $_1\mathrm{d}x^\mu$ is tangent to the 'parameter-lines' of the coordinate x^1, i.e. the curves on which only the coordinate x^1 changes. Similarly the vector $_2\mathrm{d}x^\mu$ is tangent to the parameter-lines of the coordinate x^2. Therefore, we shall have $g_{12}=0$ at a point P if and only if the parameter-lines of the coordinates x^1 and x^2 are orthogonal at P.

11. The Determinant of $g_{\mu\nu}$. Metrical Densities

We consider now the determinant of $g_{\mu\nu}$,

$$\det g_{\mu\nu} \equiv g . \tag{11.1}$$

According to (3.5) the quantity g is a scalar density of weight $w=2$:

$$\tilde{g} = g D^2 , \tag{11.2}$$

this being an immediate consequence of the transformation formula (10.5). Equation (11.2) can be written also in the form

$$\sqrt{|\tilde{g}|} = \sqrt{|g|} \cdot D , \tag{11.3}$$

which means that $\sqrt{|g|}$ is a scalar density of weight $w=1$.

It follows now from Equation (3.15) that the quantity

$$\sqrt{|g|}\, \delta V \equiv \sqrt{|g|}\, \mathrm{d}x^1\, \mathrm{d}x^2 \ldots \mathrm{d}x^n \tag{11.4}$$

is a scalar. This is the *invariant* volume element of the Riemannian space.

In general we shall have to demand that

$$g \neq 0 \quad \text{or} \quad \infty . \tag{11.5}$$

If the determinant g vanishes at a point P, we shall have the following situation. The n independent vectors:

$$(\mathrm{d}x^1, 0, \ldots, 0), (0, \mathrm{d}x^2, \ldots, 0), \ldots, (0, 0, \ldots, \mathrm{d}x^n)$$

given at P define an element of the n-dimensional space which has, according to (11.4), the invariant volume zero. Such a point P is an example of a *singular* point.

A singularity at P will be an apparent one if it is the consequence of the use of coordinates which have some defect at P. An example is given by the Euclidean 3-dimensional space. When we use Cartesian coordinates, the matrix $g_{\mu\nu}$ is according to (10.1) a diagonal one with $g=1$. But if we introduce polar coordinates r, θ, φ we shall have

$$\mathrm{d}\sigma^2 = \mathrm{d}r^2 + r^2(\mathrm{d}\theta^2 + \sin^2\theta\,\mathrm{d}\varphi^2). \tag{11.6}$$

The determinant g is now found to be

$$g = r^4 \sin^2\theta. \tag{11.7}$$

Therefore in these coordinates we shall have $g=0$ either when $r=0$ or when $\theta=0$ or π. This is evidently an apparent or a *coordinate singularity*, as it does not exist in Cartesian coordinates.

There are also singularities which are essential or *intrinsic*. We shall see an example later in the discussion of the Schwarzschild solution to the field equations of general relativity. The exact definition of an intrinsic singularity and the possibility of the existence of physically meaningful singularities, predicted by general relativity, constitute difficult problems which are still studied.

When the condition (11.5) is satisfied, we can define the *contravariant* form $g^{\mu\nu}$ of the metrical tensor by the relation

$$g^{\lambda\nu} g_{\mu\nu} = \delta^\lambda_\mu. \tag{11.8}$$

The solution of this equation is:

$$g^{\lambda\nu} = \frac{1}{g} G^{\lambda\nu}, \tag{11.9}$$

$G^{\lambda\nu}$ being the minor determinant of the matrix $g_{\lambda\mu}$ corresponding to the element $g_{\lambda\nu}$.

With $g^{\lambda\mu}$ we can now raise lower indices of tensors:

$$v^\mu = g^{\mu\nu} v_\nu, \quad T^{\mu\nu} = g^{\mu\varrho} T_\varrho^{\ \nu} = g^{\mu\varrho} g^{\nu\sigma} T_{\varrho\sigma} \quad \text{etc.} \tag{11.10}$$

The first of these equations is the inverse of Equation (10.6), i.e. the solution of (10.6) with respect to v^ν.

We now assume that in a space of n dimensions a metric has been given by the $n(n+1)/2$ functions $g_{\mu\nu}(x^\alpha)$, satisfying the condition $g \neq 0$. The following two theorems are then valid.

THEOREM 1. *It is always possible to find coordinate transformations* $x^\mu \to \tilde{x}^\mu$ *so that at any given point* P *the tensor* $\tilde{g}_{\mu\nu}$ *has vanishing non-diagonal components and each of the diagonal components is equal to* 1 *or to* -1:

$$\tilde{g}_{\mu\nu} = 0 \quad \text{if} \quad \mu \neq \nu, \qquad \tilde{g}_{\mu\nu} = \pm 1 \quad \text{if} \quad \mu = \nu. \tag{11.11}$$

The proof is obtained by the methods used in the theory of transformations of quadratic forms.

THEOREM 2. *In the diagonal form* (11.11) *the number of components equal to* $+1$ *and the number of components equal to* -1 *do not change, if we consider only real transformation* (x^μ as well as $\tilde{x}^\mu$ real). The difference of these two numbers is called the *signature* of the metric.

EXAMPLE. In Minkowski space, the metric of which is given by (10.2), we have the signature 2. The same signature will be required for the Riemannian spaces used in general relativity.

In general it is not possible to obtain the metric in the diagonal form (11.11) for all the points of any region V of the space. This is possible only in a special case, which will be characterized in Section 14.

12. The Connection of a Riemannian Space: Christoffel Symbols

In Section 7 we defined the parallel transport of a vector with respect to an arbitrary connection. We shall now prove that in a Riemannian space there is a special connection derived directly from the metrical tensor. In order to determine this connection explicitly it is sufficient to formulate the following requirement, suggested by what we know in Euclidean geometry: If a vector a^λ is given at some point P, its length must remain unchanged under parallel transport to neighbouring points P'.

The length of a contravariant vector a^λ is defined in the same way as the length of the infinitesimal vector $\mathrm{d}x^\lambda$. If we denote by $|a^\lambda|$ the length of a^λ, we shall have according to (10.3):

$$|a^\lambda|^2 = g_{\lambda\mu} a^\lambda a^\mu . \tag{12.1}$$

The requirement that the length of a^λ be unchanged under parallel transport will be therefore expressed by the following relation:

$$g_{\lambda\mu}(P)\, a^\lambda(P)\, a^\mu(P) = g_{\lambda\mu}(P')\, a^\lambda(P')\, a^\mu(P') . \tag{12.2}$$

In this formula $a^\lambda(P')$ is the vector at P' resulting from the parallel transport of $a^\lambda(P)$ to P'. If the vector **PP**$'$ has the components $\mathrm{d}x^\lambda$, we shall have:

$$g_{\lambda\mu}(P') = g_{\lambda\mu}(P) + g_{\lambda\mu,\,\nu}\, \mathrm{d}x^\nu . \tag{12.3}$$

The components $a^\lambda(P')$ are determined by Equation (7.5):

$$a^\lambda(P') = a^\lambda(P) - \Gamma^\lambda_{\mu\nu} a^\mu\, \mathrm{d}x^\nu . \tag{12.4}$$

Introducing (12.3) and (12.4) in (12.2) we find:

$$(g_{\lambda\mu,\,\nu} - g_{\alpha\mu}\Gamma^\alpha_{\lambda\nu} - g_{\lambda\alpha}\Gamma^\alpha_{\mu\nu})\, a^\lambda a^\mu\, \mathrm{d}x^\nu = 0 .$$

Since this relation must be valid for any vector a^λ and any displacement $\mathrm{d}x^\nu$ we must have:

$$g_{\lambda\mu,\,\nu} - g_{\alpha\mu}\Gamma^\alpha_{\lambda\nu} - g_{\lambda\alpha}\Gamma^\alpha_{\mu\nu} = 0 . \tag{12.5}$$

Equation (12.5) is symmetric in λ, μ and consequently it has $n^2(n+1)/2$ components. Therefore it cannot determine uniquely a general non-symmetric $\Gamma^{\lambda}_{\mu\nu}$. It has, however, exactly the number of components necessary to determine a symmetric connection. So we decide to impose on $\Gamma^{\lambda}_{\mu\nu}$ the condition

$$\Gamma^{\lambda}_{\mu\nu} = \Gamma^{\lambda}_{\nu\mu} \tag{12.6}$$

and we then determine the connection by solving Equation (12.5).

In order to solve this equation we apply to it twice the cyclic permutation of the indices $\lambda \to \mu \to \nu \to \lambda$. This gives:

$$g_{\mu\nu,\lambda} - g_{\alpha\nu}\Gamma^{\alpha}_{\mu\lambda} - g_{\mu\alpha}\Gamma^{\alpha}_{\nu\lambda} = 0 = g_{\nu\lambda,\mu} - g_{\alpha\lambda}\Gamma^{\alpha}_{\nu\mu} - g_{\nu\alpha}\Gamma^{\alpha}_{\lambda\mu}.$$

Adding the last two equations and subtracting (12.5) we find:

$$\Gamma_{\nu,\lambda\mu} = \tfrac{1}{2}(g_{\lambda\nu,\mu} + g_{\nu\mu,\lambda} - g_{\lambda\mu,\nu});\tag{12.7}$$

$$\Gamma_{\nu,\lambda\mu} \equiv g_{\nu\alpha}\Gamma^{\alpha}_{\lambda\mu}.\tag{12.8}$$

Multiplying (12.8) by $g^{\nu\varrho}$ we find, because of (11.8):

$$\Gamma^{\varrho}_{\lambda\mu} = g^{\varrho\nu}\Gamma_{\nu,\lambda\mu} = \tfrac{1}{2}g^{\varrho\nu}(g_{\nu\mu,\lambda} + g_{\lambda\nu,\mu} - g_{\lambda\mu,\nu}).\tag{12.9}$$

The quantities $\Gamma^{\varrho}_{\lambda\mu}$ given by (12.9) are called the *Christoffel symbols*. They are often denoted also by $\{{}^{\varrho}_{\lambda\mu}\}$. The connection (12.9) is used exclusively in Riemannian geometry.

Equation (12.5) can be written in an extremely simple equivalent form. Indeed, remembering the formula (6.6) for the covariant derivative of a tensor we see at once that we can write it as follows:

$$g_{\lambda\mu;\nu} = 0.\tag{12.10}$$

Thus the riemannian connection is the symmetric connection which makes the covariant derivative of $g_{\lambda\mu}$ vanish, i.e. which makes $g_{\lambda\mu}$ *covariantly constant.*

A formula similar to (12.10) is valid for the contravariant form $g^{\lambda\mu}$ of the metrical tensor. We shall determine it by taking the covariant derivative of Equation (11.8). Remembering Equations (6.7) and (12.10) we find at once:

$$g^{\alpha\mu}{}_{;\varrho}\, g_{\alpha\nu} = 0.$$

Or, if we multiply this relation by $g^{\lambda\nu}$:

$$g^{\lambda\mu}{}_{;\varrho} = 0.\tag{12.11}$$

The detailed form of this equation is, according to (6.4):

$$g^{\lambda\mu}{}_{,\varrho} + \Gamma^{\lambda}_{\alpha\varrho}g^{\alpha\mu} + \Gamma^{\mu}_{\alpha\varrho}g^{\lambda\alpha} = 0.\tag{12.12}$$

If we multiply Equation (12.5) by $g^{\lambda\mu}$ we find:

$$g^{\lambda\mu}g_{\lambda\mu,\nu} = 2\Gamma^{\lambda}_{\lambda\nu}.\tag{12.13}$$

On the other side we have from (6.10) and (11.9):

$$g_{,\nu} = g_{\lambda\mu,\nu} G^{\lambda\mu} = g_{\lambda\mu,\nu} g^{\lambda\mu} g\,. \tag{12.14}$$

Combining (12.13) and (12.14) we find:

$$g_{,\nu} - 2g\Gamma^{\lambda}_{\lambda\nu} = 0\,. \tag{12.15}$$

We can write this relation also in the form:

$$(\sqrt{|g|})_{,\nu} - \sqrt{|g|}\,\Gamma^{\lambda}_{\lambda\nu} = 0\,. \tag{12.16}$$

Since $\sqrt{|g|}$ is a scalar density of weight $w=1$, Equation (12.16) is according to (6.17) equivalent to

$$(\sqrt{|g|})_{;\nu} = 0\,. \tag{12.17}$$

The practical meaning of the Equations (12.10), (12.11) and (12.17) is that the quantities $g_{\lambda\mu}$, $g^{\lambda\mu}$ and $\sqrt{|g|}$ can in actual calculations commute with any symbol of covariant differentiation.

In Section 5 we proved a theorem on symmetric connections, according to which any given symmetric connection $\Gamma^{\lambda}_{\mu\nu}$ can be made equal to zero at a point P,

$$(\Gamma^{\lambda}_{\mu\nu})_P = 0\,,$$

through an appropriate coordinate transformation. Since the Riemannian connection is symmetric, the theorem is applicable: In a Riemannian space we can obtain

$$\left\{ {\lambda \atop \mu\nu} \right\}_P = 0 \tag{12.18}$$

through a coordinate transformation. Taking into account Equation (12.5) we see at once that we can write the Equation (12.18) also in the form

$$(g_{\lambda\mu,\nu})_P = 0\,. \tag{12.19}$$

A coordinate system, in which (12.18) or (12.19) is valid, is called *geodesic at P*. Such a coordinate system can be used in order to derive in a simple way results needing otherwise lengthy calculations. We shall give an example in Section 14. As we shall see in detail in Section 18, the result (12.18) is important also for the formulation of the principle of equivalence, which is the essential foundation of general relativity.

13. Geodesics in a Riemannian Space

In Section 8 we have defined the geodesics with respect to any given connection. This definition is applicable also to a Riemannian space, where we shall of course use the Riemannian connection.

The existence of the metric tensor $g_{\lambda\mu}$, which defines the line element ds^2 according to (10.3), suggests to use as a parameter λ on the curve the *proper length s*,

$$s = \int_{P_0}^{P} ds\,, \tag{13.1}$$

of the arc P_0P of the curve from some initial point P_0 to the point P. With $\lambda = s$ we shall have:

$$\frac{dx^\mu}{d\lambda} = \frac{dx^\mu}{ds} \equiv u^\mu . \tag{13.2}$$

The tangent vector u^μ is normalized according to (10.3) as follows:

$$g_{\lambda\mu} u^\lambda u^\mu = 1 . \tag{13.3}$$

With $\lambda = s$ the geodesic Equation (8.5),

$$\frac{du^\mu}{ds} + \Gamma^\mu_{\varrho\sigma} u^\varrho u^\sigma = f(s)\, u^\mu , \tag{13.4}$$

will be simplified. If we multiply it by

$$u_\mu = g_{\mu\alpha} u^\alpha , \tag{13.5}$$

we find:

$$f(s) = u_\mu \left(\frac{du^\mu}{ds} + \Gamma^\mu_{\varrho\sigma} u^\varrho u^\sigma \right). \tag{13.6}$$

Taking into account the relation

$$\frac{du^\mu}{ds} = u^\mu{}_{,\alpha} u^\alpha \tag{13.7}$$

we can rewrite the left-hand side of Equation (13.4) in the form:

$$(u^\mu{}_{,\sigma} + \Gamma^\mu_{\varrho\sigma} u^\varrho)\, u^\sigma = u^\mu{}_{;\sigma} u^\sigma . \tag{13.8}$$

Now we have:

$$u_\mu u^\mu{}_{;\sigma} = g_{\mu\nu} u^\nu u^\mu{}_{;\sigma} = \tfrac{1}{2} (g_{\mu\nu} u^\mu u^\nu)_{;\sigma} = 0 , \tag{13.9}$$

because of (13.3). Therefore:

$$f(s) = 0 . \tag{13.10}$$

I.e. s is an affine parameter of the geodesic, the Equation (13.4) reducing to

$$\frac{du^\mu}{ds} + \Gamma^\mu_{\varrho\sigma} u^\varrho u^\sigma = 0 . \tag{13.11}$$

Note that Equation (13.11) can be written also according to (13.8) in the form:

$$u^\mu{}_{;\nu} u^\nu = 0 . \tag{13.12}$$

The geodesics of a Riemannian space have the following important property. A geodesic connecting two points A and B is distinguished from the neighbouring lines connecting A and B as the line of minimum or maximum length. We shall give a condensed description of the proof, because it is an application of the variational method which we shall use later in the development of general relativity.

The length of the curve C connecting A and B (Figure 2) is:

$$L = \int_A^B \mathrm{d}s = \int_A^B \{g_{\mu\nu}\dot{x}^\mu \dot{x}^\nu\}^{1/2}\, \mathrm{d}s\,; \qquad \dot{x}^\mu = \frac{\mathrm{d}x^\mu}{\mathrm{d}s}. \tag{13.13}$$

The neighbouring curve $\tilde{C}$ connecting A and B will be described by the equations

$$\tilde{x}^\mu = x^\mu + \varepsilon\xi^\mu, \tag{13.14}$$

the vector ξ^μ satisfying the condition

$$\xi^\mu = 0 \quad \text{at} \quad A \text{ and } B. \tag{13.15}$$

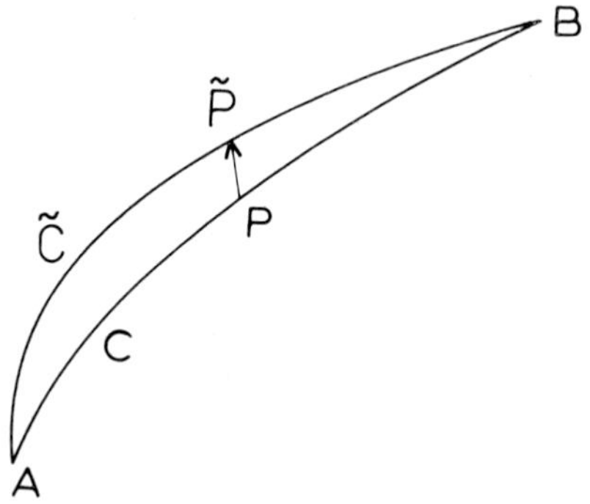

Fig. 2. Variation of a curve connecting two points.

The length of $\tilde{C}$ will be given by the formula

$$\tilde{L} = \int_A^B \{g_{\mu\nu}(\tilde{x}^\alpha)\, \dot{\tilde{x}}^\mu \dot{\tilde{x}}^\nu\}^{1/2}\, \mathrm{d}s\,; \qquad \dot{\tilde{x}}^\mu = \frac{\mathrm{d}\tilde{x}^\mu}{\mathrm{d}s} = \dot{x}^\mu + \varepsilon\dot{\xi}^\mu, \tag{13.16}$$

in which we use as parameter the proper length of C. Let us write:

$$\{g_{\mu\nu}\dot{x}^\mu \dot{x}^\nu\}^{1/2} = \varphi(x^\alpha, \dot{x}^\alpha)\,; \qquad \{g_{\mu\nu}(\tilde{x}^\alpha)\, \dot{\tilde{x}}^\mu \dot{\tilde{x}}^\nu\}^{1/2} = \tilde{\varphi}(\tilde{x}^\alpha, \dot{\tilde{x}}^\alpha). \tag{13.17}$$

From (13.13), (13.16) and (13.17) we have:

$$\delta L \equiv \tilde{L} - L = \int_A^B \delta\varphi\, \mathrm{d}s\,; \tag{13.18}$$

$$\delta\varphi \equiv \tilde{\varphi} - \varphi = \varepsilon\left\{\frac{\partial\varphi}{\partial x^\alpha} - \frac{\mathrm{d}}{\mathrm{d}s}\left(\frac{\partial\varphi}{\partial \dot{x}^\alpha}\right)\right\}\xi^\alpha + \varepsilon\,\frac{\mathrm{d}}{\mathrm{d}s}\left(\frac{\partial\varphi}{\partial \dot{x}^\alpha}\,\xi^\alpha\right). \tag{13.19}$$

The last term in (13.19) does not contribute to δL because of (13.15). The condition that the length of C be an extremum will therefore be expressed by the relation:

$$\delta L = \varepsilon\int_A^B \left\{\frac{\partial\varphi}{\partial x^\alpha} - \frac{\mathrm{d}}{\mathrm{d}s}\left(\frac{\partial\varphi}{\partial \dot{x}^\alpha}\right)\right\}\xi^\alpha\, \mathrm{d}s = 0\,. \tag{13.20}$$

Since ξ^α is arbitrary, we must have at each point of C:

$$\frac{\partial\varphi}{\partial x^\alpha} - \frac{\mathrm{d}}{\mathrm{d}s}\left(\frac{\partial\varphi}{\partial\dot{x}^\alpha}\right) = 0. \tag{13.21}$$

This is the differential equation of the curve C which satisfies the condition (13.20).

We shall show that Equation (13.21) is just another form of the geodesic Equation (13.12). From (13.17) we find at once:

$$\frac{\partial\varphi}{\partial x^\alpha} = \frac{\partial\varphi}{\partial g_{\mu\nu}}\, g_{\mu\nu,\,\alpha} = \frac{1}{2\varphi}\, g_{\mu\nu,\,\alpha}\dot{x}^\mu\dot{x}^\nu\,; \qquad \frac{\partial\varphi}{\partial\dot{x}^\alpha} = \frac{1}{\varphi}\, g_{\mu\alpha}\dot{x}^\mu\,. \tag{13.22}$$

We note that on C we have according to (10.3) $\varphi = 1$. Introducing the relations (13.22) into (13.21) we obtain the equation:

$$\tfrac{1}{2} g_{\mu\nu,\,\alpha} u^\mu u^\nu - \frac{\mathrm{d}u_\alpha}{\mathrm{d}s} = 0. \tag{13.23}$$

We have used in this relation the notation (13.2). With the help of (5.10) and (12.9) one verifies that Equation (13.23) is equivalent to

$$u_{\alpha;\,\beta} u^\beta = 0. \tag{13.24}$$

This is the covariant form – lower index α – of the geodesic equation, which is obtained by multiplying the contravariant form (13.12) by $g_{\alpha\mu}$. Inversely, we get from (13.24) the equation (13.12) if we multiply (13.24) by $g^{\mu\alpha}$.

To end this discussion we have to remark that in Riemannian spaces with indefinite metric, as are the spaces used in general relativity, the definition (13.2) of u^μ as well as the derivation of Equation (13.21) are possible only for lines which are not null. On any null-line the proper length vanishes and consequently it cannot be used as a parameter for the detailed description of the line.

14. The Curvature of a Riemannian Space: The Riemann Tensor

We shall apply the general definition of the curvature tensor, derived in Section 9 from any given connection, using now the Riemannian connection. Since the Riemannian connection is symmetric, we shall have for the *Riemann tensor* $R^\varrho{}_{\lambda\mu\nu}$ not only the general symmetry property (9.4) but also (9.8):

$$R^\varrho{}_{\lambda\mu\nu} = -R^\varrho{}_{\lambda\nu\mu}\,, \qquad R^\varrho{}_{[\lambda\mu\nu]} = 0\,. \tag{14.1}$$

Since we have now a metric, we can lower the index ϱ and we shall then find additional symmetries of the curvature tensor. We have from (9.3):

$$\begin{aligned} R_{\alpha\lambda\mu\nu} &= g_{\alpha\varrho} R^\varrho{}_{\lambda\mu\nu} = g_{\alpha\varrho}\left(-\Gamma^\varrho_{\lambda\mu,\,\nu} + \Gamma^\varrho_{\lambda\nu,\,\mu}\right) + \cdots \\ &= -\left(g_{\alpha\varrho}\Gamma^\varrho_{\lambda\mu}\right)_{,\,\nu} + \left(g_{\alpha\varrho}\Gamma^\varrho_{\lambda\nu}\right)_{,\,\mu} + \cdots, \end{aligned}$$

the omitted terms being quadratic in the Christoffel symbols. Using (12.7) we write this relation in the form:

$$R_{\alpha\lambda\mu\nu} = \tfrac{1}{2}(g_{\lambda\mu,\alpha\nu} + g_{\alpha\nu,\lambda\mu} - g_{\alpha\mu,\lambda\nu} - g_{\lambda\nu,\alpha\mu}) + \cdots . \tag{14.2}$$

In a geodesic coordinate system all omitted terms vanish and we can obtain the additional symmetries of the Riemann tensor from the expression written in detail in (14.2). They are the following:

$$R_{\alpha\lambda\mu\nu} = -R_{\lambda\alpha\mu\nu} = R_{\mu\nu\alpha\lambda} . \tag{14.3}$$

We consider now the case $n=4$ of general relativity and we shall determine the number of independent components of the Riemann tensor in this case. It will be convenient to use the concept of a *bivector* and of the *bivector space.* An antisymmetric tensor of order 2, $F_{\lambda\mu} = -F_{\mu\lambda}$, is called a bivector. In the space of 4 dimensions the bivector has 6 independent components, e.g. the components corresponding to the values of the indices

$$\lambda\mu = 01, 02, 03;\ 23, 31, 12.$$

(In general relativity one uses often the notation x^0 instead of x^4.) We shall now write $F_{\lambda\mu}$ as a formal vector F_A, the index A taking the values 1, 2,..., 6, and we shall say that F_A is a vector in the bivector space which has 6 dimensions. We have to introduce some correspondence between the values of the indices A and $\lambda\mu$. We choose the following:

$$\begin{Bmatrix} \lambda\mu = 01, & 02, & 03, & 23, & 31, & 12 \\ A = \ \ 1, & 2, & 3, & 4, & 5, & 6 \end{Bmatrix}. \tag{14.4}$$

A tensor T_{AB} of order 2 in the bivector space is actually a tensor $T_{\lambda\mu\nu\varrho}$ of order 4 in the Riemannian space, which is antisymmetric in λ, μ as well as in ν, ϱ. Therefore the Riemann tensor, having the symmetry properties (14.3) and the first of (14.1), will correspond to a symmetric tensor of order 2 in the bivector space:

$$R_{\lambda\mu\nu\varrho} \leftrightarrow R_{AB} = R_{BA} . \tag{14.5}$$

A symmetric tensor of order 2 in a space of 6 dimensions has $\frac{1}{2}\cdot 6\cdot 7 = 21$ independent components. Actually the number of the independent components of the Riemann tensor will be smaller, because we have to consider also the second of Equations (14.1) or equivalently the equation

$$R_{\alpha[\lambda\mu\nu]} = 0 . \tag{14.6}$$

In this relation the indices λ, μ, ν must evidently have different values. It can be verified easily that, if α also differs from λ, μ, ν, we have relations which are all equivalent to the following one condition:

$$3R_{0[123]} = R_{0123} + R_{0231} + R_{0312} = 0 . \tag{14.6$'$}$$

In the bivector space this condition reads, according to (14.4):

$$R_{14} + R_{25} + R_{36} = 0 . \tag{14.7}$$

If α has the same value as one of the indices λ, μ, ν, Equation (14.6) simply repeats some of Equations (14.3). Therefore Equation (14.6) gives only the one condition (14.7) and consequently the Riemann tensor of the 4-dimensional space has $21-1=20$ independent components.

The contraction (9.10) of the Riemann tensor vanishes identically,

$$R^{\lambda}{}_{\lambda\mu\nu} = g^{\lambda\alpha} R_{\alpha\lambda\mu\nu} = 0\,,$$

because of (14.3) and (10.4). The only non-vanishing contraction is therefore the *Ricci tensor*:

$$R_{\mu\nu} = R^{\lambda}{}_{\mu\nu\lambda}\,. \tag{14.8}$$

Since

$$R_{\mu\nu} = g^{\lambda\alpha} R_{\alpha\mu\nu\lambda} = g^{\lambda\alpha} R_{\nu\lambda\alpha\mu} = g^{\lambda\alpha} R_{\lambda\nu\mu\alpha}\,,$$

we see that the Ricci tensor is symmetric:

$$R_{\mu\nu} = R_{\nu\mu}\,. \tag{14.9}$$

Therefore in a 4-dimensional riemannian space the Ricci tensor has 10 independent components. With the metric tensor $g^{\mu\nu}$ we can contract also the Ricci tensor:

$$g^{\mu\nu} R_{\mu\nu} = R^{\nu}{}_{\nu} = R\,, \tag{14.10}$$

R being called the *curvature scalar*. The combination

$$R_{\mu\nu} - \tfrac{1}{2} g_{\mu\nu} R \equiv G_{\mu\nu} \tag{14.11}$$

is called the *Einstein tensor*, as it is the tensor appearing in the field equations of general relativity.

The Riemann tensor satisfies the following important identity:

$$R^{\lambda}{}_{\mu[\nu\varrho;\sigma]} = 0\,. \tag{14.12}$$

This is the *Bianchi identity*, which can be verified easily from (14.2), with the help of a geodesic coordinate system. The contraction $\lambda=\sigma$ of (14.12) is:

$$R^{\lambda}{}_{\mu\nu\varrho;\lambda} + R_{\mu\varrho;\nu} - R_{\mu\nu;\varrho} = 0\,. \tag{14.13}$$

It can be shown that (14.13) is equivalent to (14.12), because of the symmetries of the Riemann tensor. We can contract again, multiplying (14.13) by $g^{\mu\varrho}$. The result is the following identity:

$$(R^{\lambda}{}_{\nu} - \tfrac{1}{2}\delta^{\lambda}{}_{\nu} R)_{;\lambda} = 0\,, \tag{14.14}$$

expressing the fact that the covariant divergence of the Einstein tensor vanishes. All of these identities play an important role in general relativity.

Another important relation is obtained when we try to express the Riemann tensor in terms of trace-free tensor quantities. The final result of this discussion is the following relation:

$$\left.\begin{aligned} R_{\gamma\mu\nu\varrho} = C_{\lambda\mu\nu\varrho} &+ \tfrac{1}{2}(g_{\lambda\varrho}B_{\mu\nu} + g_{\mu\nu}B_{\lambda\varrho} - g_{\lambda\nu}B_{\mu\varrho} - g_{\mu\varrho}B_{\lambda\nu}) \\ &+ \tfrac{1}{12}R(g_{\lambda\varrho}g_{\mu\nu} - g_{\lambda\nu}g_{\mu\varrho})\,. \end{aligned}\right\} \tag{14.15}$$

The tensor $B_{\mu\nu}$ is given by the equation

$$B_{\mu\nu} = R_{\mu\nu} - \tfrac{1}{4} g_{\mu\nu} R, \tag{14.16}$$

while the tensor $C^{\lambda}{}_{\mu\nu\varrho}$ is defined by the Equation (14.15). It is verified at once that the tensor $B_{\mu\nu}$ is traceless:

$$g^{\mu\nu} B_{\mu\nu} = 0. \tag{14.16$'$}$$

If we multiply (14.15) by $g^{\lambda\varrho}$, we find finally:

$$g^{\lambda\varrho} C_{\lambda\mu\nu\varrho} = 0, \tag{14.17}$$

i.e. the tensor $C_{\lambda\mu\nu\varrho}$ is also traceless.

The tensor $C_{\lambda\mu\nu\varrho}$ is the *Weyl tensor*. It is called also the *conformal curvature tensor* because it has the following property. Consider besides the Riemannian space S with the metric $g_{\mu\nu}$ a second Riemannian space $\tilde{S}$ having the metric

$$\tilde{g}_{\mu\nu} = \varphi g_{\mu\nu}. \tag{14.18}$$

The space $\tilde{S}$ is said to be conformal to S. The Riemann tensor $\tilde{R}^{\varrho}{}_{\lambda\mu\nu}$ of $\tilde{S}$ will in general differ from $R^{\varrho}{}_{\lambda\mu\nu}$. One can, however, prove the following relation:

$$\tilde{C}^{\varrho}{}_{\lambda\mu\nu} = C^{\varrho}{}_{\lambda\mu\nu}. \tag{14.19}$$

I.e. the 'conformal transformation' (14.18) does not change the Weyl tensor $C^{\varrho}{}_{\lambda\mu\nu}$.

It can be verified at once from (14.15) that the Weyl tensor has the same symmetries as the Riemann tensor:

$$C_{\varrho\lambda\mu\nu} = -C_{\lambda\varrho\mu\nu} = -C_{\varrho\lambda\nu\mu} = C_{\mu\nu\varrho\lambda}; \qquad C_{\varrho[\lambda\mu\nu]} = 0. \tag{14.20}$$

These symmetry properties will leave as many independent components of $C_{\lambda\mu\nu\varrho}$ as there are independent components of $R_{\lambda\mu\nu\varrho}$, i.e. 20. We have still to consider the condition (14.17). This is an equation symmetric in μ, ν and consequently it contains 10 independent relations between the $C_{\lambda\mu\nu\varrho}$. Therefore the Weyl tensor has $20-10=10$ independent components.

The theorem established in Section 9 for any symmetric connection will be valid in the Riemannian space for the Riemannian connection. Therefore, in a Riemannian space the necessary and sufficient condition for the parallel transport to be path-independent is the vanishing of the Riemann tensor:

$$R^{\varrho}{}_{\lambda\mu\nu} = 0. \tag{14.21}$$

In this case we can introduce coordinates such that

$$\Gamma^{\lambda}_{\mu\nu} = 0 \quad \text{or} \quad g_{\lambda\mu,\nu} = 0 \tag{14.22}$$

everywhere. Moreover we can in this case put the metric $g_{\mu\nu}$ in diagonal form, with its diagonal components equal to ± 1 in the whole space. A space having these properties is called a *flat space*.

15 Algebraic Classification of the Weyl Tensor

In a metric space we can obtain an invariant classification of a tensor $T_{\alpha\beta}$ of order 2 by the method of eigenvectors and eigenvalues. The method is based on the following equation:

$$T_{\alpha\beta}V^{\beta} = \lambda V_{\alpha} = \lambda g_{\alpha\beta}V^{\beta}, \tag{15.1}$$

λ being a scalar and V^{β} a vector. A vector $V^{\beta} \neq 0$, which satisfies Equation (15.1), is called an *eigenvector* of $T_{\alpha\beta}$. If we write Equation (15.1) in the form

$$(T_{\alpha\beta} - \lambda g_{\alpha\beta})\, V^{\beta} = 0, \tag{15.2}$$

we see at once that an eigenvector can exist only for the values of λ, which are roots of the equation:

$$\det(T_{\alpha\beta} - \lambda g_{\alpha\beta}) = 0. \tag{15.3}$$

These values of λ are the *eigenvalues* of $T_{\alpha\beta}$.

The discussion of Equation (15.2) is a *local* problem, referring to some point P of the space. Consequently we can simplify the equation by using a frame in which $g_{\alpha\beta}$ is diagonal, with its diagonal components equal to ± 1. We are interested in the Riemannian spaces used in general relativity, which are 4-dimensional and of signature -2. We therefore shall write $g_{\alpha\beta}$ in the form:

$$g_{\alpha\beta} = \begin{pmatrix} 1 & 0 & 0 & 0 \\ 0 & -1 & 0 & 0 \\ 0 & 0 & -1 & 0 \\ 0 & 0 & 0 & -1 \end{pmatrix}. \tag{15.4}$$

As a first example of the application of this method we mention the discussion of the electromagnetic field-tensor $F_{\alpha\beta}$. In the case of the gravitational field it is the Weyl tensor of the Riemannian space, whose classification is important. The Weyl tensor being of order 4, it cannot be discussed directly according to the Equation (15.2). However, we have remarked in Section 14 that a tensor of order 4 having the symmetry properties of the Riemann tensor can be considered as a symmetric tensor of order 2 in the 6-dimensional bivector space. Since the Weyl tensor has all these symmetries, we can write:

$$C_{\alpha\beta\mu\nu} \leftrightarrow C_{AB} = C_{BA}\,; \qquad A, B = 1, 2, \ldots, 6.$$

In order to write an equation of the form (15.2) for C_{AB} we need a 'metric' g_{AB} in the bivector space. The corresponding tensor $g_{\mu\nu\varrho\sigma}$ in the Riemannian space can be constructed from the metrical tensor $g_{\mu\nu}$ in a unique way. It is the tensor:

$$g_{\mu\nu\varrho\sigma} = g_{\mu\varrho}g_{\nu\sigma} - g_{\mu\sigma}g_{\nu\varrho}. \tag{15.5}$$

One verifies at once that this tensor has the same symmetries as the Weyl tensor. Consequently the corresponding tensor g_{AB} in the bivector space is symmetric in A, B.

The equation of the form (15.2) for the Weyl tensor will now be:

$$(C_{AB} - \lambda g_{AB})\, W^B = 0\,. \tag{15.6}$$

Note that this equation is the transcription for the bivector space of the following equation in the Riemannian space:

$$(C_{\mu\nu\varrho\sigma} - \lambda g_{\mu\nu\varrho\sigma})\, F^{\varrho\sigma} = 0\,,$$

$F^{\varrho\sigma}$ being an antisymmetric tensor. The use of the bivector space simplifies the discussion and so we shall work with Equation (15.6).

For the correspondence between the bivector indices A and the pairs of indices $\mu\nu$ in the Riemannian space we adopt the relation (14.4). We find then at once from (15.5) that with the values (15.4) of $g_{\mu\nu}$ the g_{AB} is also diagonal:

$$\left.\begin{array}{l} g_{AB} = 0 \quad \text{if} \quad A \neq B\,; \\ \text{diagonal components} \quad g_{AB} = (-1, -1, -1; 1, 1, 1)\,. \end{array}\right\} \tag{15.7}$$

In order to determine the general form of the matrix C_{AB} we have to transcribe in bivector form all the relations satisfied by the Weyl tensor. An example of this transcription has been given in Section 14, where we derived the bivector form of the second of Equations (14.1) for the Riemann tensor. Since the Weyl tensor also satisfies this equation, we shall have, in analogy with (14.7):

$$C_{14} + C_{25} + C_{36} = 0\,. \tag{15.8}$$

A similar transcription has to be made for all other relations satisfied by the Weyl tensor, i.e. for Equations (14.17) and (14.20). The final result obtained from the discussion of the resulting equations is the following: The matrix C_{AB} has the form

$$(C_{AB}) = \begin{pmatrix} (M) & (N) \\ (N) & (-M) \end{pmatrix}, \tag{15.9}$$

(M) and (N) being two symmetric 3-dimensional matrices with vanishing trace:

$$(M) \equiv (m_{ik}) = (m_{ki})\,, \qquad (N) \equiv (n_{ik}) = (n_{ki})\,; \qquad i, k = 1, 2, 3\,; \tag{15.10}$$

$$m_{11} + m_{22} + m_{33} = n_{11} + n_{22} + n_{33} = 0\,. \tag{15.11}$$

Because of (15.11) each of the matrices (M) and (N) has $6-1=5$ independent elements. Consequently the matrix C_{AB} has 10 independent elements, in accordance with the fact that the Weyl tensor $C_{\mu\nu\varrho\sigma}$ has 10 independent components.

The eigenvalues λ of Equation (15.6) are the roots of the equation

$$\det(C_{AB} - \lambda g_{AB}) = 0\,. \tag{15.12}$$

This 6-dimensional determinant can be reduced to a 3-dimensional one, because of the symmetry of the matrix C_{AB}. We add to the first column of the determinant (15.12) the second column multiplied by i and then we add to the second row the first multiplied by i. The result is:

$$\begin{vmatrix} m_{ik} + \lambda\delta_{ik} + \mathrm{in}_{ik} & n_{ik} \\ 0 & -(m_{ik} + \lambda\delta_{ik} - in_{ik}) \end{vmatrix} = 0 .$$

This determinant is the product of the determinants $|m_{ik}+\lambda\delta_{ik}+in_{ik}|$ and $-|m_{ik}+ +\lambda\delta_{ik}-in_{ik}|$. Thus Equation (15.12) reduces to two equations with 3-dimensional matrices. The first equation is

$$|m_{ik} + in_{ik} + \lambda\delta_{ik}| = 0 , \tag{15.13}$$

the second being the complex-conjugate to (15.13).

If we write (15.13) in detail, we find at once that it is a cubic equation for λ of the form:

$$\lambda^3 + \lambda(...) + (...) = 0 . \tag{15.14}$$

The coefficient of λ^2 is equal to zero because of (15.11). Therefore Equation (15.13) has 3 roots λ_1, λ_2 and λ_3 with vanishing sum:

$$\lambda_1 + \lambda_2 + \lambda_3 = 0 . \tag{15.15}$$

Equation (15.12) has the roots λ_1, λ_2, λ_3 and their complex-conjugates $\bar{\lambda}_1, \bar{\lambda}_2, \bar{\lambda}_3$.

The *Petrov classification* of the Weyl tensor is based on the type of the solution to the Equation (15.13). We shall have one of the following 3 types:

(i) the 3 roots λ_1, λ_2, λ_3 are all different. In this case we say that the Weyl tensor is of the Petrov type I.

(ii) Two roots are equal, $\lambda_1 = \lambda_2 \neq \lambda_3$. The Weyl tensor is then called of type II.

(iii) The 3 roots are equal and according to (15.15)

$$\lambda_1 = \lambda_2 = \lambda_3 = 0 .$$

The Weyl tensor is then of type III.

When the eigenvalues are known, the corresponding eigenvectors have to be determined from Equation (15.6). This equation can also be reduced to a 3-dimensional one. If we write the 6 components W^A as follows:

$$W^\alpha = (W^i, \tilde{W}^k) ; \qquad i, k = 1, 2, 3 , \tag{15.16}$$

Equation (15.6) separates into the following two equations:

$$\left.\begin{aligned} (m_{ik} + \lambda\delta_{ik})\, W^k + n_{ik}\tilde{W}^k = 0 , \\ n_{ik}W^k - (m_{ik} + \lambda\delta_{ik})\, \tilde{W}^k = 0 . \end{aligned}\right\} \tag{15.17}$$

If we add the second of these equations, multiplied by i, to the first we find:

$$(m_{ik} + in_{ik} + \lambda\delta_{ik})\,(W^k - i\tilde{W}^k) = 0 . \tag{15.18}$$

This 3-dimensional equation contains the two Equations (15.17) and is therefore equivalent to the 6-dimensional Equation (15.6).

When we know the eigenvectors, which correspond to the different Petrov types, we can determine the *normal form* in which we can put the matrix C_{AB}, i.e. the 3-dimensional matrices M and N. We shall not give the details of this discussion but only the final results.

A Weyl tensor of the type I has the following normal form:

$$\left.\begin{aligned}&(M)=\begin{pmatrix}\alpha_1 & 0 & 0\\ 0 & \alpha_2 & 0\\ 0 & 0 & \alpha_3\end{pmatrix}, \qquad (N)=\begin{pmatrix}\beta_1 & 0 & 0\\ 0 & \beta_2 & 0\\ 0 & 0 & \beta_3\end{pmatrix};\\ &\alpha_1+\alpha_2+\alpha_3=\beta_1+\beta_2+\beta_3=0.\end{aligned}\right\} \tag{15.19}$$

The 6 eigenvalues are:

$$\lambda_1=-(\alpha_1+i\beta_1), \qquad \lambda_2=-(\alpha_2+i\beta_2), \qquad \lambda_3=-(\alpha_3+i\beta_3)$$

and the complex-conjugates $\bar{\lambda}_1$, $\bar{\lambda}_2$, $\bar{\lambda}_3$.

In the case of the type II the normal form is:

$$(M)=\begin{pmatrix}2\alpha & 0 & 0\\ 0 & -\alpha+\sigma & 0\\ 0 & 0 & -\alpha-\sigma\end{pmatrix}, \qquad N=\begin{pmatrix}2\beta & 0 & 0\\ 0 & -\beta & \sigma\\ 0 & \sigma & -\beta\end{pmatrix} \tag{15.20}$$

The single root is $\lambda_1=-2(\alpha+i\beta)$ and the double root $\lambda_2=\lambda_3=\alpha+i\beta$.

For the type III the normal form obtained by Petrov is:

$$(M)=\begin{pmatrix}0 & \sigma & 0\\ \sigma & 0 & 0\\ 0 & 0 & 0\end{pmatrix}, \qquad (N)=\begin{pmatrix}0 & 0 & 0\\ 0 & 0 & \sigma\\ 0 & \sigma & 0\end{pmatrix}. \tag{15.21}$$

A more detailed classification of the Weyl tensor can be obtained if one starts from the definition of a *principal null vector* k^α of $C_{\mu\nu\varrho\sigma}$. This is a vector satisfying the following *equation of Debever-Penrose*:

$$k_{[\varrho}C_{\alpha]\beta\gamma[\delta}k_{\sigma]}k^\beta k^\gamma=0. \tag{15.22}$$

It can be proved that this equation has 4 solutions k^α.

The following 5 cases are now possible:

(1) The 4 solutions of (15.22) are all different (simple roots only).

(2) There is one double and two simple roots of (15.22).

(3) Two double roots.

(4) One triple and one simple root.

(5) All 4 roots are equal.

The detailed discussion shows that in the case (1) we have the Petrov type I. In the case (2) the Weyl tensor is of type II. In the case (3) the Weyl tensor is said to be of *type D*: This is the special case of the Petrov type II corresponding to $\sigma=0$. In the case (4) the Weyl tensor is of type III. Finally in the case (5) the Weyl tensor is called of *type N*. This is a special case of type III having the following normal form:

$$(M)=\begin{pmatrix}0 & 0 & 0\\ 0 & \sigma & 0\\ 0 & 0 & -\sigma\end{pmatrix}, \qquad (N)=\begin{pmatrix}0 & 0 & 0\\ 0 & 0 & \sigma\\ 0 & \sigma & 0\end{pmatrix}. \tag{15.23}$$

The type N is characterized by the existence of a null-vector k^σ satisfying the relation

$$C_{\mu\nu\varrho\sigma}k^\sigma = 0\,, \tag{15.24}$$

k^σ being the unique solution of Equation (15.22). Similarly the type III is characterized by the existence of a null-vector k^σ satisfying the relation

$$C_{\mu\nu\varrho[\sigma}k_{\alpha]}k^\varrho = 0\,, \tag{15.25}$$

k^σ being the triple root of (15.22). More complicated relations characterize the cases (3) and (2).

The type I is called also *algebraically general*, all other types being *algebraically special.*

We only mention here, without entering into any details, that a very simple and elegant derivation of the detailed classification of the Weyl tensor is obtained using the *spinor description* of the gravitational field.

16. Lie Derivatives. Isometries, Killing Vectors

A coordinate transformation $x^\mu \to \tilde{x}^\mu$ has been given until now the following meaning: To the point P, which has initially the coordinate values x^μ, we assign new coordinate values $\tilde{x}^\mu$ determined from x^μ with the help of the n given functions

$$\tilde{x}^\mu = f^\mu(x^\alpha)\,. \tag{16.1}$$

We shall now give to the transformation $x^\mu \to \tilde{x}^\mu$ the following entirely different meaning: To the point P having the coordinate values x^μ we make correspond another point Q of the same space having the coordinate values $\tilde{x}^\mu$ *in the same coordinate system.* This operation is called a *mapping of the space onto itself.*

We shall consider the following *infinitesimal mapping*:

$$\tilde{x}^\mu = x^\mu + \varepsilon\xi^\mu\,. \tag{16.2}$$

In this formula ξ^μ is some given vectorfield, $\xi^\mu = \xi^\mu(x^\alpha)$, and ε an infinitesimal parameter. The meaning of Equation (16.2) is that to the point P with coordinate values x^μ we make correspond the point Q with coordinate values $x^\mu + \varepsilon\xi^\mu$ (in the same coordinate system).

Let us now assume that some tensor field has been given in the space. We have then at Q first the tensor determined by the given field at Q. A second tensor at the point Q will result from the tensor determined by the given field at P when this tensor is 'taken over' to the point Q in a manner corresponding to the mapping (16.2). Thus we shall have two tensors at Q. Comparing them we shall be in a position to define the *Lie derivative*, with respect to the mapping (16.2), of the given tensor field.

We consider first a scalar field $\varphi(x^\mu)$. It has at the point P the value $\varphi_P = \varphi(x^\mu)$. This being a scalar, we postulate that in the mapping $P \to Q$ we have to take over to Q the value φ_P unchanged. We shall then have at Q the two scalars:

$$\varphi_Q = \varphi(\tilde{x}^\mu) = \varphi_P + \varepsilon\varphi_{,\mu}\xi^\mu \quad \text{and} \quad \varphi_{P\to Q} = \varphi_P\,. \tag{16.3}$$

The Lie derivative of the scalar φ with respect to ξ^μ is defined as follows:

$$\underset{\xi}{\pounds}\varphi = \lim_{\varepsilon\to 0} \frac{\varphi_Q - \varphi_{P\to Q}}{\varepsilon}. \tag{16.4}$$

Therefore from (16.3):

$$\underset{\xi}{\pounds}\varphi = \varphi_{,\mu}\xi^\mu. \tag{16.5}$$

Let us now consider a contravariant vector field $k^\mu = k^\mu(x^\alpha)$. In this case we shall find what will be the result of the operation $k^\mu_{P\to Q}$ in the following way. The vector k^μ may be thought as the tangent vector at P to a curve passing through P. If this curve is described by the scalar parameter λ we shall have:

$$k^\mu = \frac{\mathrm{d}x^\mu}{\mathrm{d}\lambda}, \tag{16.6}$$

$\mathrm{d}x^\mu$ being the components of the vector **PP**$'$ (Figure 3). Let the points Q and Q' correspond to P and P' by the mapping (16.2). The components δx^μ of the vector **QQ**$'$ will be:

$$\delta x^\mu = \mathrm{d}x^\mu + \varepsilon\xi^\mu_{P'} - \varepsilon\xi^\mu_P = \mathrm{d}x^\mu + \varepsilon\xi^\mu{}_{,\nu}\mathrm{d}x^\nu. \tag{16.7}$$

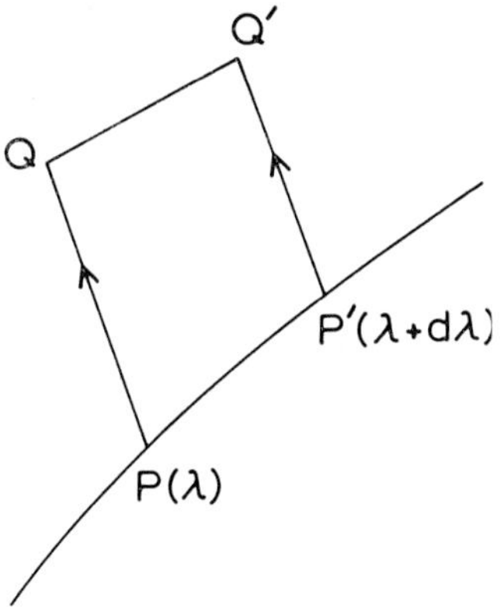

Fig. 3. Determining the Lie derivative of a contravariant vector.

Since the parameter λ is a scalar, the result of the transport of the vector k^μ from P to Q by the mapping (16.2) will be:

$$k^\mu_{P\to Q} = \frac{\delta x^\mu}{\mathrm{d}\lambda} = k^\mu_P + \varepsilon\xi^\mu{}_{,\nu}k^\nu. \tag{16.8}$$

The definition of the Lie derivative of the vector k^μ is:

$$\underset{\xi}{\pounds}k^\mu = \lim_{\varepsilon\to 0} \frac{k^\mu_Q - k^\mu_{P\to Q}}{\varepsilon}. \tag{16.9}$$

Since

$$k^\mu_Q = k^\mu(\tilde{x}^\alpha) = k^\mu_P + \varepsilon k^\mu{}_{,\nu}\xi^\nu$$

we find finally:

$$\underset{\xi}{\pounds} k^\mu = k^\mu{}_{,\nu}\xi^\nu - \xi^\mu{}_{,\nu}k^\nu . \tag{16.10}$$

The Lie derivative of any other tensor $T^{\cdot\cdot}_{\cdot\cdot}$ will be defined as in (16.4) or (16.9):

$$\underset{\xi}{\pounds} T^{\cdot\cdot}_{\cdot\cdot} = \lim_{\varepsilon\to 0} \frac{(T^{\cdot\cdot}_{\cdot\cdot})_Q - (T^{\cdot\cdot}_{\cdot\cdot})_{P\to Q}}{\varepsilon} . \tag{16.11}$$

The detailed formula for any given tensor will be derived from the results obtained for the scalar and the contravariant vector, combined with the requirement of a rule of the type (6.2) for the Lie derivative of a product of tensors. Thus we find for a covariant vectorfield p_μ with the help of an arbitrary contravariant vector k^μ:

$$\underset{\xi}{\pounds}(p_\mu k^\mu) = \underset{\xi}{\pounds} p_\mu \cdot k^\mu + p_\mu \underset{\xi}{\pounds} k^\mu . \tag{16.12}$$

For the left-hand side we have from (16.5):

$$\underset{\xi}{\pounds}(p_\mu k^\mu) = (p_\mu k^\mu)_{,\nu}\,\xi^\nu = (p_{\mu,\nu} k^\mu + p_\mu k^\mu{}_{,\nu})\,\xi^\nu .$$

Introducing in the last term of (16.12) the expression (16.10) we find:

$$k^\mu \left\{ \underset{\xi}{\pounds} p_\mu - p_{\mu,\nu}\xi^\nu - p_\nu \xi^\nu{}_{,\mu} \right\} = 0 .$$

Since k^μ is arbitrary, we shall have:

$$\underset{\xi}{\pounds} p_\mu = p_{\mu,\nu}\xi^\nu + p_\nu \xi^\nu{}_{,\mu} . \tag{16.13}$$

By similar calculations we derive the formula for the Lie derivative af any tensor. We give here the formula for a covariant tensor $T_{\mu\nu}$:

$$\underset{\xi}{\pounds} T_{\mu\nu} = T_{\mu\nu,\varrho}\xi^\varrho + T_{\varrho\nu}\xi^\varrho{}_{,\mu} + T_{\mu\varrho}\xi^\varrho{}_{,\nu} . \tag{16.14}$$

Applied to the metrical tensor this formula gives:

$$\underset{\xi}{\pounds} g_{\mu\nu} = g_{\mu\nu,\varrho}\xi^\varrho + g_{\varrho\nu}\xi^\varrho{}_{,\mu} + g_{\mu\varrho}\xi^\varrho{}_{,\nu} . \tag{16.15}$$

If we replace $g_{\mu\nu,\varrho}$ by its value derived from (12.5) we can rewrite this equation as follows:

$$\underset{\xi}{\pounds} g_{\mu\nu} = g_{\varrho\nu}\xi^\varrho{}_{;\mu} + g_{\mu\varrho}\xi^\varrho{}_{;\nu} .$$

Since

$$g_{\varrho\nu}\xi^\varrho{}_{;\mu} = (g_{\varrho\nu}\xi^\varrho)_{;\mu} = \xi_{\nu;\mu} ,$$

we can write simply:

$$\underset{\xi}{\pounds} g_{\mu\nu} = \xi_{\mu;\nu} + \xi_{\nu;\mu} . \tag{16.16}$$

We shall now ask the following question: If the neighbouring points P and P' go over, under the mapping (16.2), to the points Q and Q' (Figure 3), what is the condition ensuring that

$$\mathrm{d}s^2_{\mathbf{PP'}} = \mathrm{d}s^2_{\mathbf{QQ'}} \tag{16.17}$$

for any pair of points P, P'? If such a mapping exists, it will be called an *isometric mapping*.

The answer to this question is straightforward. We have

$$\mathrm{d}s^2_{\mathbf{PP'}} = (g_{\mu\nu})_P\, \mathrm{d}x^\mu\, \mathrm{d}x^\nu,$$

$\mathrm{d}x^\mu$ being the components of $\mathbf{PP'}$ and $(g_{\mu\nu})_P$ the metric tensor at P. But again, according to the second of (16.3):

$$\mathrm{d}s^2_{\mathbf{PP'}} = (g_{\mu\nu}\, \mathrm{d}x^\mu\, \mathrm{d}x^\nu)_{P\to Q} = (g_{\mu\nu})_{P\to Q}\cdot \delta x^\mu\, \delta x^\nu,$$

δx^μ being the components of $\mathbf{QQ'}$. On the other side we have:

$$\mathrm{d}s^2_{\mathbf{QQ'}} = (g_{\mu\nu})_Q\, \delta x^\mu \delta x^\nu.$$

Therefore the condition for the validity of (16.17) is:

$$(g_{\mu\nu})_{P\to Q} = (g_{\mu\nu})_Q\,.$$

I.e., according to the definition (16.11) of the Lie derivative:

$$\underset{\xi}{\pounds} g_{\mu\nu} = 0\,. \tag{16.18}$$

We have thus found that the condition for the existence of isometric mappings is the existence of solutions ξ^μ of the equation

$$\xi_{\mu;\,\nu} + \xi_{\nu;\,\mu} = 0\,. \tag{16.19}$$

This is the *Killing equation* and vectors ξ^μ which satisfy it are called *Killing vectors*.

The existence of a Killing vector, i.e. of an isometric mapping of the space onto itself, is the expression of a certain intrinsic symmetry property of the space. The following example is important in general relativity. A *stationary gravitational field* is a Riemannian space having a Killing vector ξ^μ which is time-like, $\xi^\mu\xi_\mu > 0$. Let us consider the trajectories of the field ξ^μ. We can then construct a coordinate system in such a way that on these trajectories only one coordinate, e.g. the coordinate x^0, changes. Then the trajectories of ξ^μ are the parameter lines of x^0 and we shall have in this coordinate system:

$$\xi^\alpha = 0 \quad \text{if} \quad \alpha \neq 0\,.$$

Moreover we can obtain

$$\xi^0 = 1\,.$$

We now derive at once from (16.18) with the help of (16.15):

$$g_{\mu\nu,\,0} = 0: \tag{16.20}$$

In such a coordinate system the components of the metric are independent of the coordinate x^0. We say in this case that the coordinate system has been *adapted* to the stationary character of the metric. The use of such coordinates is extremely useful, when one tries to find stationary solutions of the Einstein equations, because it simplifies greatly the otherwise much too complicated field equations.

More examples of Killing vectors will be given later.

Exercises

III1: In a 4-dimensional space verify the relation:

$$g \equiv \det g_{\lambda\mu} = \frac{1}{4!}\,\varepsilon^{\lambda\mu\nu\varrho}\varepsilon^{\alpha\beta\gamma\delta} g_{\lambda\alpha} g_{\mu\beta} g_{\nu\gamma} g_{\varrho\delta}\,.$$

Use this result to prove (12.17).

III2: Prove that in a 4-dimensional space the relation (14.6) is equivalent to:

$$R_{\lambda\mu\nu\varrho} = R_{\nu\varrho\lambda\mu} \quad \text{and} \quad R_{1[234]} = 0\,.$$

III3: Verify by direct computation the relations (15.9), (15.10) and (15.11).

III4: Discuss the eigenvalue problem for a tensor $F_{\lambda\mu} = F_{[\lambda\mu]}$ (electromagnetic field tensor).

III5: Derive the formula for the Lie derivative of a scalar density. (In the case of weight $w=1$ start from Equation (3.1′).)

III6: Verify that the formula for the Lie derivative of any tensor can be rewritten with all ordinary derivatives replaced by covariant ones.

CHAPTER IV

THE GRAVITATIONAL FIELD

17. Introductory Remarks

We mentioned already in the Introduction that the Newtonian theory of gravitation has been extremely successful in the detailed study of the motion of planets. Nevertheless, this theory cannot be maintained as it is, because it uses the prerelativistic concepts of absolute space and absolute time: We have to modify it at least so as to make use of the space-time of the special theory of relativity. Because of the successes of the newtonian theory we shall have to make on the new theory the following two demands:

(1) In an appropriate first approximation for weak gravitational fields the new theory must reduce to the Newtonian one, giving again the same results for the planetary motions.

(2) Beyond this approximation the new theory may predict small deviations from the predictions of the Newtonian theory. These deviations will have to be subsequently verified by the observations.

The characteristic properties of the Newtonian theory are the following:

(1) It is a *scalar* theory; it has a scalar potential φ and, correspondingly, a scalar source of the field, this source being the mass-density of the material distribution.

(2) The field equation is a linear differential equation of the second order:

$$\Delta\varphi = -\,4\pi G\mu\,. \tag{17.1}$$

Δ is the Laplace operator and G the gravitational constant entering into Newton's law for gravitational attraction:

$$F = \frac{Gmm'}{r^2}\,. \tag{17.2}$$

The simplest relativistic generalization of this theory would be a theory using again a scalar gravitational potential. In this case the source of the field must be some scalar quantity derived from the relativistic description of matter.

Continuous matter is described in special relativity by the *matter tensor*, also known as the *stress-energy-momentum tensor*. This is a symmetric tensor $T_{\mu\nu}$,

$$T_{\mu\nu} = T_{\nu\mu}\,. \tag{17.3}$$

The physical significance of the components $T_{\mu\nu}$ can be understood by taking as an example the Maxwell stress-energy-momentum tensor ${}_{\mathrm{M}}T_{\mu\nu}$ of the electromagnetic field. Indeed, if in a region of the space containing matter there is also an electromagnetic field, the situation will be described by the sum $T_{\mu\nu} + {}_{M}T_{\mu\nu}$ and consequently the significance of the components of these two tensors must be the same.

The component ${}_{M}T_{00}$ of the Maxwell tensor is the electromagnetic energy density. Consequently T_{00} will be the material energy density. The component ${}_{M}T_{0i}$ $(i=1, 2, 3)$ is either the component of the momentum density of the electromagnetic field in the direction of the axis Ox^i (multiplied by c) or the flow of electromagnetic energy in the direction Ox^i. Therefore T_{0i} will have the same meaning for the material distribution. Finally the components ${}_{M}T_{ik}$ $(i, k=1, 2, 3)$ are the electromagnetic stresses. Therefore T_{ik} will be the mechanical stresses acting in the continuous matter, which we are considering. In more detail the component T_{11} is the normal stress on a surface-element orthogonal to Ox^1; T_{12} is the shearing stress in the direction Ox^2 on a surface element orthogonal to Ox^1, or the stress in the direction Ox^1 on a surface element orthogonal to Ox^2, and so on.

In special relativity wa have the relation

$$E = mc^2 \tag{17.4}$$

between the energy and the mass of a particle. Since T_{00} is the energy density, we could define the mass density by the relation

$$\mu = \frac{1}{c^2} T_{00} .$$

This is, however, just one component of a tensor and not a scalar. Consequently it cannot be used as the source of the field in a scalar theory of gravitational interaction. There is only one scalar which can be constructed from $T_{\mu\nu}$ and this is the contraction of $T_{\mu\nu}$ with the metric tensor or, in other words, the trace of $T_{\mu\nu}$:

$$T = \eta^{\mu\nu} T_{\mu\nu} = T^{\nu}_{\nu} , \tag{17.5}$$

$\eta^{\mu\nu}$ being the metric tensor of Minkowski space. The scalar T corresponds qualitatively to the density of proper mass.

A theory using the scalar T as the source of the gravitational field was proposed by Nordström a few years before the formulation of general relativity. This is a truly relativistic theory, formulated in the Minkowski space-time of special relativity. The detailed discussion of this theory has shown that it satisfies the first condition, i.e. it reduces to the Newtonian theory of gravitation in the first approximation. However, in the next approximation it leads to results which are in disagreement with the observations. First, it predicts a rotation of the perihelion of Mercury which has not only a wrong value, but even the wrong sign. It also predicts a zero value for the deflection of light, this being an immediate consequence of the fact that the trace of the Maxwell tensor ${}_{M}T_{\mu\nu}$ vanishes.

It is not possible to construct a vector from the tensor $T_{\mu\nu}$ in a straightforward way. It follows that a vectorial theory of the gravitational field, in which the source as well as the potential are vectors, cannot be constructed but only with the help of complicated and artificial assumptions. Thus the next possibility, which we have to consider, is the *tensor* theory of the gravitational field. In such a theory all 10 components of the tensor $T_{\mu\nu}$ will act as sources of the gravitational field. Correspondingly the potential will also be a symmetric tensor of order 2.

The simplest idea would be to formulate such a theory in the Minkowski space of special relativity. Theories of this type have been considered often in the past and are still being considered. They are somehow mathematically simpler than general relativity, but they meet with important conceptual difficulties. Another significant difficulty is the one of choosing a theory from the existing great variety of mathematically possible theories of this type. This contrasts with the situation in general relativity, where the field equations are almost uniquely determined, as we shall see in Section 19.

General relativity is also a tensorial theory of gravitation, with the tensor $T_{\mu\nu}$ acting as the source of the gravitational field. However, the flat Minkowski space of special relativity is replaced in general relativity by a *curved Riemannian space*. The metric tensor $g_{\mu\nu}$ of this riemannian space will play the role of the gravitational potential. It is the tensor $g_{\mu\nu}$ which describes the gravitational field in a way which we shall explain in detail in the following sections.

18. The Principle of Equivalence

Einstein arrived at the radical decision to develop a theory of the gravitational interaction based on a curved Riemannian space by formulating and then using systematically the *equivalence principle*. This is a generalization of the following assumption, found to be so successful in the Newtonian theory of gravitation: The 'gravitational mass' of a body, entering into the law (17.2) of gravitational attraction, is the same as the 'inertial' mass, entering into the law of motion

$$\mathbf{F} = m\gamma\,. \tag{18.1}$$

An indirect verification of this assumption is given by the successes of the Newtonian theory of gravitation, combined with Newtonian mechanics, in solving the problem of planetary motions. As a direct verification we mention the experiments of Eőtvős. In these experiments the gravitational force, which is proportional to the gravitational mass, is compared to the centrifugal force due to the rotation of the Earth, this second force being proportional to the inertial mass. The result of the experiments of Eőtvős is:

$$(m_{\mathrm{gr}} - m_{\mathrm{in}})/m_{\mathrm{gr}} < 10^{-8}\,.$$

Recent improvement of this result has been obtained by Dicke, the new limit being:

$$(m_{\mathrm{gr}} - m_{\mathrm{in}})/m_{\mathrm{gr}} < 10^{-11}\,.$$

Einstein generalized this assumption into the following *principle of equivalence*: Gravitational and inertial forces are completely equivalent from the physical point of view; i.e. they are of an identical nature and consequently *it is impossible to separate them by any physical experiment*. The so formulated principle proved to be very powerful indeed, as it will be seen from the following consequences drawn from it.

(1) Gravitational forces, or equivalently gravitational accelerations, will have to be described in the same way as inertial forces, or inertial accelerations. It is very easy to see how we describe inertial accelerations. In an inertial frame in Minkowski space a free particle moves on a straight line with the equation

$$\frac{d^2x^\mu}{ds^2} = 0. \tag{18.2}$$

This is the geodesic Equation (13.11) simplified by the fact that in an inertial frame the components of the metric are constants and consequently the Christoffel symbols, given by (12.9), vanish. If now we introduce general coordinates, the equation of motion will be the general form of the equation of geodesics:

$$\frac{d^2x^\mu}{ds^2} + \Gamma^\mu_{\varrho\sigma}\frac{dx^\varrho}{ds}\frac{dx^\sigma}{ds} = 0. \tag{18.3}$$

The second term in this equation is the inertial acceleration of the particle, the existence of which is a consequence of the fact that we are now using a non-inertial frame. Thus we see that the inertial accelerations are described quite generally by the Christoffel symbols. It follows now from the principle of equivalence that gravitational accelerations must be described also by the Christoffel symbols. Since the Christoffel symbols are derived from the metric tensor, we conclude further that the metric tensor will play the role of the gravitational potential. More exactly, the Christoffel symbols will now describe, according to the principle of equivalence, the *sum* of the inertial and the gravitational acceleration. Moreover, there will be no possibility to split this sum unambiguously by any physical experiment into two terms representing separately the inertial and the gravitational acceleration.

(2) When there are gravitational accelerations present, as for example in the gravitational field of the earth, the space cannot be the flat Minkowski space. Indeed, in the Minkowski space we can have

$$\Gamma^\lambda_{\mu\nu} = 0$$

everywhere. This should then be interpreted as meaning that the sum of the inertial and the gravitational acceleration could be made equal to zero everywhere. This does, however, not correspond to our experience about gravitational accelerations: When gravitational accelerations exist, it is not possible to make them vanish everywhere. We can only make them vanish at one point, or approximately in a small region, by the use of an appropriate coordinate system. Therefore, when a gravitational field is present, the space will be necessarily a *curved* Riemannian space. The gravitational field will then appear as the expression of the fact that we are in a curved riemannian space and no longer in the flat Minkowski space. We describe this result by saying that in general relativity the gravitational field has been *reduced to the geometry* or, in other words, that the gravitational field has been *geometrized.*

(3) An immediate consequence of the preceding remarks will be the non-existence of inertial frames in general relativity. Indeed, an inertial frame is by definition one in which the inertial accelerations vanish and consequently the gravitational accelerations would have been separated from the inertial ones, in contradiction to the equivalence principle. Only in the case of weak gravitational fields will it be possible to reintroduce the inertial frames of special relativity as a first approximation, as we shall see in detail in Section 31.

The non-existence of inertial frames means that in general relativity the acceleration also ceases to have an absolute meaning. We may recall that the special theory of relativity postulates that the velocity has a relative meaning only; but the acceleration continues to have there an absolute meaning. It is because of this fact that the new theory is called general relativity.

The numerous consequences drawn from the principle of equivalence are not only logically consistent, but also extremely well adapted to the situations encountered in physical problems. In order to illustrate this statement we give the following two examples.

In a Riemannian space we can always make the Christoffel symbols vanish on a given curve by an appropriate coordinate transformation. According to the principle of equivalence this property of the Riemannian space should mean physically that the sum of the inertial and the gravitational acceleration can be made equal to zero on the given curve. This is really the case, as we see at once if we consider a freely falling lift or a freely moving non-rotating artificial satellite: An observer inside the satellite, using the frame which is connected rigidly with the satellite – a comoving frame –, will observe neither inertial non gravitational accelerations.

As a second example we shall describe the reasoning which was used by Einstein in order to show the necessity of the *deflection of a light ray* in a gravitational field. Consider an observer at rest in a lift and a light ray passing in his neighbourhood. If the lift is not accelerated, the observer will find that the ray moves on the straight line 1 (Figure 4). If now the lift starts being accelerated, e.g. upwards, the observer will see the light ray moving, relativ to a frame connected rigidly with the lift, on the curved line 2. This is due to the fact that in the frame connected with the lift there is now an inertial acceleration directed downwards. According to the principle of equivalence the same must happen if the lift is not accelerated, but is situated in a gravitational field with the gravitational acceleration directed downwards. Thus *a light ray has to be deflected in a gravitational field*, a result which has been verified subsequently by observations. We shall treat this deflection quantitatively in Section 24.

Another important consequence of the equivalence principle is the following. Consider a particle which moves under the action of the gravitational field only. Let L be the trajectory of the particle and $\tilde{x}^\alpha$ the coordinates in which the Christoffel symbols $\tilde{\Gamma}^\lambda_{\mu\nu}$ vanish along L. In this frame the sum of the inertial and the gravitational acceleration acting upon the particle is equal to zero. Since by hypothesis there is no other force of any kind acting on the particle, its total acceleration will be zero, i.e. we shall have the equation of motion

$$\frac{d^2\tilde{x}^\alpha}{ds^2} = 0. \tag{18.4}$$

This is the geodesic equation simplified by the fact that the $\tilde{\Gamma}^\lambda_{\mu\nu}$ vanish on L. In another frame x^α we shall have as equation of motion of the particle the geodesic Equation (13.11):

$$\frac{du^\mu}{ds} + \Gamma^\mu_{\varrho\sigma} u^\varrho u^\sigma = 0. \tag{18.5}$$

This is the *law of geodesic motion*, valid for particles on which the gravitational field alone is acting.

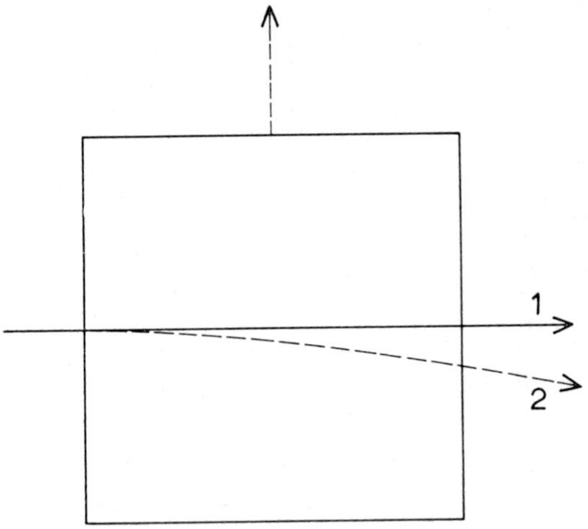

Fig. 4. Deflection of a light ray.

It must however be added that the Equation (18.5) is valid only for the motion of particles which can be considered as *test particles.* A test particle is by definition a small particle moving in the neighbourhood of bodies of much larger masses. In this case we neglect the contribution of the test particle to the field, i.e. we assume that we have the field created by the other bodies in the absence of the test particle. When, on the contrary, we are considering bodies of comparable masses, as for example a double star, the gravitational field will be changing significantly in the interior of each body because of the contribution to the total field of the body itself. Consequently it is not possible to find out what part of the body, if some part at all, will move on a geodesic.

The situation was clarified later, when Einstein proved that in general relativity the equations of motion of bodies interacting through their gravitational field cannot be prescribed arbitrarily, because they are determined by the field equations. The derivation of the equations of motion for a given type of body constitutes a very difficult problem, which has been solved only approximately in some simple cases. We shall describe this problem in some detail in Chapter X.

Here we shall only remark that one of the simple problems which have been solved is the motion of test particles and that the discussion of the motion of the test particles of the simplest type (particles without spin) has led to the geodesic law (18.5). This result illustrates the importance of the equivalence principle and, more generally, the remarkable coherence of the ideas on which general relativity is based.

19. The Field Equations of General Relativity

Since the matter tensor $T_{\mu\nu}$ is the source of the gravitational field and $g_{\mu\nu}$ is its potential, the field equations must be of the form

$$F_{\mu\nu} = T_{\mu\nu}, \tag{19.1}$$

the expression $F_{\mu\nu}$ being constructed from $g_{\mu\nu}$ and its derivatives. In order to maintain the qualitative analogy with the newtonian theory of gravitation we shall allow in the expression $F_{\mu\nu}$ derivatives of $g_{\mu\nu}$ only up to the second order and we shall demand that $F_{\mu\nu}$ be linear in the second derivatives of $g_{\mu\nu}$. Furthermore, since $T_{\mu\nu}$ is a tensor, we shall have to demand that $F_{\mu\nu}$ be also a tensor. Thus the field equations will form a *tensor equation*, having the same form in all coordinate systems. Note that in the present case writing the field equations in a form which is valid in all frames is indispensable, as there are no preferred frames in general relativity.

The field Equations (19.1) will determine, when solved, the metric $g_{\mu\nu}$ of the space from the given distribution of the sources $T_{\mu\nu}$. We see that we are in an entirely new situation: The metric is no longer given *a priori* but has to be determined from the material distribution which we are considering. In a more general statement we may say that the metric of the space as a whole is determined by the distribution of matter in the Universe. This is one of the formulations of *Mach's principle*, which played an important heuristic role in Einstein's efforts to arrive at the detailed formulation of general relativity.

The field equations can be derived from a variational principle. This derivation will be discussed in Section 33. The equations can be derived also in a direct, elementary way which we shall describe now.

A theorem in Riemannian geometry states that the only tensors which can be constructed from the $g_{\mu\nu}$ and their first and second derivatives and which are linear in the second derivatives are the Riemann tensor and its contractions, and of course also the tensor $g_{\mu\nu}$ itself. Therefore the general form of $F_{\mu\nu}$ will be:

$$F_{\mu\nu} = \alpha R_{\mu\nu} + \beta g_{\mu\nu} R + \gamma g_{\mu\nu}, \tag{19.2}$$

α, β and γ being constants. The first term of this expression cannot vanish, because $T_{\mu\nu}$ cannot be proportional to $g_{\mu\nu}$ except in very special cases. Consequently we must have $\alpha \neq 0$. We can then divide Equation (19.1) by α and so obtain the form:

$$R_{\mu\nu} + \tilde{\beta} g_{\mu\nu} R + \tilde{\gamma} g_{\mu\nu} = -\kappa T_{\mu\nu}. \tag{19.3}$$

The new constant $\kappa = -1/\alpha$ is the *relativistic gravitational constant*. We shall see later that κ is proportional to the gravitational constant G appearing in Equation (17.2).

The simplest field equations are obtained from (19.3) if we put $\tilde{\beta} = \tilde{\gamma} = 0$:

$$R_{\mu\nu} = -\kappa T_{\mu\nu}. \tag{19.4'}$$

These are the equations which were postulated by Einstein initially. In regions of space containing no matter we have $T_{\mu\nu} = 0$ and consequently Equation (19.4′) reduces to

$$R_{\mu\nu} = 0. \tag{19.4}$$

Equations (19.4) are the correct equations of the gravitational field for *empty space*, containing no matter at all. On the contrary, it has been found necessary to modify Equation (19.4′) because of the following difficulty. The matter tensor obeys in special relativity, when we use inertial frames, the conservation law:

$$T^{\mu\nu}{}_{,\nu} = 0.$$

If we allow general coordinates, this relation has to be written in the form:

$$T^{\mu\nu}{}_{;\nu} = 0. \tag{19.5}$$

According to the principle of equivalence we have to demand the validity of Equation (19.5) also in the Riemannian space which describes the gravitational field in general relativity. But then it follows from (19.4′) that the Ricci tensor must satisfy the relation

$$R^{\mu\nu}{}_{;\nu} = 0. \tag{19.6'}$$

These are 4 additional equations, besides the 10 equations (19.4′), for the 10 components $g_{\mu\nu}$, leading to an overdetermination of $g_{\mu\nu}$. However, it was noticed that because of the identity (14.14):

$$(R^{\mu\nu} - \tfrac{1}{2}g^{\mu\nu}R)_{;\nu} = 0, \tag{19.6}$$

there will be no additional conditions imposed on $g_{\mu\nu}$, if we replace Equation (19.4′) by the following equations:

$$R_{\mu\nu} - \tfrac{1}{2}g_{\mu\nu}R \equiv G_{\mu\nu} = -\kappa T_{\mu\nu}. \tag{19.7}$$

These are the correct field equations, identical with those derived from the variational principle as we shall see in Section 33. We recall the detailed expressions of $R_{\mu\nu}$ and R in terms of $g_{\mu\nu}$:

$$\left.\begin{aligned} R_{\mu\nu} &= R^{\alpha}{}_{\mu\nu\alpha} = -\Gamma^{\alpha}_{\mu\nu,\alpha} + \Gamma^{\alpha}_{\mu\alpha,\nu} - \Gamma^{\beta}_{\mu\nu}\Gamma^{\alpha}_{\beta\alpha} + \Gamma^{\beta}_{\mu\alpha}\Gamma^{\alpha}_{\beta\nu}; \\ R &= g^{\mu\nu}R_{\mu\nu}; \\ \Gamma^{\alpha}_{\mu\nu} &= \tfrac{1}{2}g^{\alpha\beta}(g_{\beta\nu,\mu} + g_{\mu\beta,\nu} - g_{\mu\nu,\beta}). \end{aligned}\right\} \tag{19.8}$$

In empty space we have $T_{\mu\nu} = 0$ and consequently Equations (19.7) reduce to

$$R_{\mu\nu} - \tfrac{1}{2}g_{\mu\nu}R = 0. \tag{19.9}$$

Multiplying this relation by $g^{\mu\nu}$ we find from (14.10) and (11.8):

$$R - \tfrac{1}{2}\delta^\mu_\mu R = 0. \tag{19.10}$$

In the case of a space of n dimensions we have $\delta^\mu_\mu = n$. Since in general relativity we have $n=4$, Equation (19.10) gives

$$R = 0$$

and consequently Equation (19.9) reduces again to (19.4).

A first, qualitative information about the constant κ will be obtained in the following way. Let us consider a coordinate system in which all four coordinates have the dimension of a length (no polar coordinates). We conclude then from (10.3) and (11.8) that the $g_{\mu\nu}$ and $g^{\mu\nu}$ are dimensionless. Consequently we shall have from (12.9) and (9.3):

$$\Gamma^\lambda_{\mu\nu} \sim l^{-1}, \quad R_{\mu\nu} \sim l^{-2}.$$

The components $T_{\mu\nu}$ have the dimensions of an energy density:

$$T_{\mu\nu} \sim mc^2 l^{-3},$$

m being some mass. Therefore from (19.7):

$$\kappa \sim l m^{-1} c^{-2}. \tag{19.11}$$

Now from Newton's law (17.2) we find, if we remember that the product rF has the dimensions of energy or of mc^2:

$$G \sim l m^{-1} c^2. \tag{19.12}$$

Comparing (19.11) and (19.12) we see that we must have

$$\kappa = \alpha \frac{G}{c^4},$$

α being some dimensionless factor. The value of α will be derived in Section 31, where we shall discuss the Equation (19.7) in the first approximation for weak gravitational fields. Demanding that in this approximation general relativity give exactly the same results as the newtonian theory of gravitation we shall find:

$$\kappa = \frac{8\pi G}{c^4}. \tag{19.13}$$

We end this section with a short discussion of the more general form of the field equations containing also the term $\tilde{\gamma} g_{\mu\nu}$ of (19.3). It is customary to write these equations in the form

$$R_{\mu\nu} - \tfrac{1}{2} g_{\mu\nu} R + \Lambda g_{\mu\nu} = -\kappa T_{\mu\nu}. \tag{19.14}$$

A comparison of the term $\Lambda g_{\mu\nu}$ with the preceding terms shows at once that the constant Λ must have the dimension l^{-2}. The discussion of Equation (19.14) for the case of a material system consisting of two bodies has shown that, when $\Lambda>0$, the term $\Lambda g_{\mu\nu}$ describes a repulsive force between the two bodies which varies as the square of the distance of the bodies. Since an interaction of this kind is not observed, we conclude that Λ is either strictly zero or so small that the term $\Lambda g_{\mu\nu}$ in (19.14) has negligible consequences in the planetary motions. We may mention that the term $\Lambda g_{\mu\nu}$ was introduced in the field equations by Einstein because it allows the existence of a *static cosmological solution* of the equations. Accordingly Einstein assumed that the length $a=\Lambda^{-1/2}$ is of the order of magnitude of the 'radius of the Universe':

$$a = \Lambda^{-1/2} \approx 10^{10}\text{pc} \approx 10^{28}\ \text{cm},$$

Λ having then an extremely small value of the order $10^{-56}\ \text{cm}^{-2}$. For this reason the term $\Lambda g_{\mu\nu}$ is called the *cosmological term* of the field equations, Λ being the *cosmological constant*.

The cosmological problem will be discussed briefly in Chapter XII. In all physical problems, in which the system under discussion is not the whole Universe, we shall put $\Lambda=0$, i.e. we shall use the Equation (19.7).

CHAPTER V

THE SCHWARZSCHILD SOLUTION

20. Metrics with Spherical Symmetry

As a first application of the field equations (19.7) we shall determine the gravitational field of a spherically symmetric body at rest. We assume that there are no contributions to the field from any other matter. Consequently the field will be spherically symmetric and time-independent or *static*.

In this section we shall determine the general form of a metric $g_{\mu\nu}$ with spherical symmetry. This can be done with the help of the Killing equation (16.18). Let us use Cartesian-like coordinates x^i in the 3-dimensional space ($i=1, 2, 3$), the origin being at the center of symmetry O. The spherical symmetry around the origin O is equivalent to a rotation symmetry about each one of the three coordinate axes. The symmetry of rotation of the metric around the axis Ox^1 is seen at once to be expressed by the existence of the following Killing vector:

$$\xi^0 = \xi^1 = 0, \qquad \xi^2 = \alpha x^3, \qquad \xi^3 = -\alpha x^2; \qquad \alpha = \text{const.}$$

The symmetry about each one of the axes Ox^i will be described by a Killing vector depending on 3 arbitrary constants α, β and γ:

$$\xi^0 = 0, \qquad \xi^i = \varepsilon^{ik}x^k; \qquad \varepsilon^{ik} = -\varepsilon^{ki} = \begin{pmatrix} 0 & \gamma & -\beta \\ -\gamma & 0 & \alpha \\ \beta & -\alpha & 0 \end{pmatrix}. \tag{20.1}$$

Introducing this vector in the Killing equation (16.18) and (16.15) we find, for example for $\mu=\nu=0$:

$$g_{00,i}\varepsilon^{ik}x^k = 0. \tag{20.2}$$

Since α, β and γ are arbitrary, Equation (20.2) splits into the 3 equations:

$$g_{00,i}x^k - g_{00,k}x^i = 0 \qquad (i, k = 1, 2, 3). \tag{20.3}$$

These equations imply that g_{00} will be an arbitrary function of the euclidean distance r from the origin:

$$r = \{(x^1)^2 + (x^2)^2 + (x^3)^2\}^{1/2}. \tag{20.4}$$

Of course, g_{00} may depend in an arbitrary way on the time t, as there are no restrictions imposed on this dependence by the Equations (20.3). Therefore the final result is that g_{00} will be an arbitrary function of the two variables r and t. Discussing in a similar way the other components of the Killing equation we determine the general form of the spherically symmetric $g_{\mu\nu}$ completely.

A more direct derivation will be obtained on the basis of the remark that there are only the following two differential expressions in x^i which are invariant under the 3-dimensional rotations:

$$(\mathrm{d}x^1)^2 + (\mathrm{d}x^2)^2 + (\mathrm{d}x^3)^2 = \mathrm{d}r^2 + r^2(\mathrm{d}\theta^2 + \sin^2\theta\,\mathrm{d}\varphi^2), \tag{20.5}$$

$$x^1\,\mathrm{d}x^1 + x^2\,\mathrm{d}x^2 + x^3\,\mathrm{d}x^3 = r\,\mathrm{d}r. \tag{20.6}$$

The angles θ and φ are defined by the equations:

$$x^1 = r\sin\theta\cos\varphi, \qquad x^2 = r\sin\theta\sin\varphi, \qquad x^3 = r\cos\theta, \tag{20.7}$$

which determine the transformation from the polar coordinates r, θ, φ to the coordinates x^i. Remembering that, from the point of view of the 3-dimensional rotations, the time t is an independent scalar variable, we conclude that the final expression for $\mathrm{d}s^2$ must be constructed from (20.5), (20.6) and $\mathrm{d}t$, multiplied by scalar coefficients.

A spherically symmetric scalar φ has to satisfy the equation

$$\underset{\xi}{\pounds}\varphi = \varphi_{,\alpha}\xi^\alpha = \varphi_{,i}\varepsilon^{ik}x^k = 0. \tag{20.8}$$

This equation is of the form of (20.2). Consequently the scalar φ must depend on r and t only, $\varphi = \varphi(r, t)$. Thus we find:

$$\left.\begin{aligned} \mathrm{d}s^2 = {} & A(r,t)\,\{(\mathrm{d}x^1)^2 + (\mathrm{d}x^2)^2 + (\mathrm{d}x^3)^2\} + \\ & + B(r,t)\,(x^1\,\mathrm{d}x^1 + x^2\,\mathrm{d}x^2 + x^3\,\mathrm{d}x^3)^2 + \Gamma(r,t)\,\mathrm{d}t^2 + \\ & + \Delta(r,t)\,(x^1\,\mathrm{d}x^1 + x^2\,\mathrm{d}x^2 + x^3\,\mathrm{d}x^3)\,\mathrm{d}t. \end{aligned}\right\} \tag{20.9}$$

We see that the general spherically symmetric metric contains 4 arbitrary functions of r and t.

It will be convenient to use systematically the polar coordinates r, θ, φ defined by (20.7). We have then, according to (20.5) and (20.6):

$$\left.\begin{aligned} \mathrm{d}s^2 = {} & \alpha(r,t)\,\mathrm{d}r^2 + \beta(r,t)\,(\mathrm{d}\theta^2 + \sin^2\theta\,\mathrm{d}\varphi^2) + \\ & + \gamma(r,t)\,\mathrm{d}t^2 + \delta(r,t)\,\mathrm{d}r\,\mathrm{d}t. \end{aligned}\right\} \tag{20.10}$$

We shall now show that we can obtain certain simplifications of the form (20.10) in a general way, i.e. without imposing any restrictions on the metric. We start by remarking that the metric (20.10) is invariant under the following two types of transformations:

(1) Transformations $(\theta, \varphi) \to (\tilde{\theta}, \tilde{\varphi})$ leaving invariant the expression $\mathrm{d}\theta^2 + \sin^2\theta\,\mathrm{d}\varphi^2$, i.e. the transformations on the 2-dimensional sphere. We shall not need these transformations explicitly.

(2) Arbitrary transformations $(r, t) \to (\tilde{r}, \tilde{t})$. It is with the help of transformations of this type that we shall be able to simplify (20.10).

We first assume that $\beta(r, t) \neq \text{const}$ and, if $\beta_{,r} \neq 0$, we apply the transformation

$$\beta(r, t) = -\tilde{r}^2, \qquad t = \tilde{t}.$$

In this way we obtain, if we write again r and t instead of $\tilde{r}$ and $\tilde{t}$, the form (20.10) with only 3 arbitrary functions α, γ and δ, the function β being replaced by $-r^2$.

This new form of ds^2 does not contain the case $\beta = \text{const}$. However, in this case also the expression for ds^2 contains only the arbitrary functions α, γ and δ. Moreover this case is physically uninteresting for the following reason.

The gravitational field must tend to zero when the distance from the field-producing body tends to an infinite value, $r \to \infty$. But when there is no gravitational field, the space is the flat Minkowski space. Therefore we must demand, when we examine any physical problem dealing with a bounded material distribution (but not the cosmological problem), that the metric $g_{\mu\nu}$ tend to the Minkowski metric when $r \to \infty$:

$$ds^2 \to (ds^2)_{\text{Mink}} \quad \text{for} \quad r \to \infty. \tag{20.11}$$

Now the Minkowski metric has in polar coordinates the form

$$(ds^2)_{\text{Mink}} = c^2\, dt^2 - dr^2 - r^2(d\theta^2 + \sin^2\theta\, d\varphi^2). \tag{20.12}$$

Consequently, if we have the form (20.10) with $\beta = \text{const}$, it will be impossible to satisfy the demand (20.11). Thus we see that the case $\beta = \text{const}$ is physically uninteresting.

Now we perform a second transformation of the form

$$r = \tilde{r}, \qquad t = \varphi(\tilde{r}, \tilde{t}).$$

We must have $\partial t/\partial \tilde{t} \neq 0$, since otherwise the transformation would be singular. The inverse transformation $(\tilde{r}, \tilde{t}) \to (r, t)$ will be:

$$\tilde{r} = r, \qquad \tilde{t} = \phi(r, t). \tag{20.13}$$

Now we find:

$$d\tilde{r} = dr, \qquad d\tilde{t} = \phi_{,r}\, dr + \phi_{,t}\, dt \quad \text{with} \quad \phi_{,t} \neq 0.$$

Therefore:

$$dr = d\tilde{r}, \qquad dt = (d\tilde{t} - \phi_{,r}\, d\tilde{r})/\phi_{,t}.$$

Introducing these relations in the expression for ds^2 we find that the product $d\tilde{r}\, d\tilde{t}$ has now the coefficient

$$(\phi_{,t}\delta - 2\phi_{,r}\gamma)(\phi_{,t})^{-2}.$$

The equation

$$\phi_{,t}\delta - 2\phi_{,r}\gamma = 0 \tag{20.14}$$

can always be solved, for any given $\gamma \neq 0$ and δ. Therefore we can make the coefficient of $d\tilde{r}\, d\tilde{t}$ equal to zero, by taking as function ϕ in (20.13) a solution of Equation (20.14). Thus we obtain a simplified form of (20.10) having $\beta = -r^2$ and $\delta = 0$.

We shall write this simplified form of ds^2 as follows:

$$ds^2 = e^{\nu} c^2\, dt^2 - e^{\mu}\, dr^2 - r^2 (d\theta^2 + \sin^2\theta\, d\varphi^2) \tag{20.15}$$

with the new arbitrary functions $\nu(t, r)$ and $\mu(t, r)$. The exponentials will be convenient for the subsequent calculations. The choice of the signs in (20.15) corresponds to the demand that the metric have the same signature as the Minkowski metric. We add the remark that the factor c^2 in the coefficient of dt^2 in (20.15) is usually omitted in the intermediate calculations.

21. The Schwarzschild Solution. Theorem of Birkhoff

In the problem which we are going to discuss now and which is to determine the gravitational field of a spherically symmetric body at rest we have to demand also that the metric be time-independent,

$$g_{\mu\nu,0} = 0 . \tag{21.1}$$

Therefore in this case the metric will be of the form (20.15) with the functions μ and ν depending on the coordinate r only. Note that in this case the metric has a fourth Killing vector which has been already discussed at the end of Section 16.

We have to start by establishing the detailed form of the field equations corresponding to the metric (20.15). This metric is diagonal. If we write:

$$t = x^0 ; \qquad r = x^1 , \qquad \theta = x^2 , \quad \varphi = x^3 , \tag{21.2}$$

we see from (20.15) that the diagonal components of the metric are:

$$g_{00} = e^{\nu} ; \qquad g_{11} = -\, e^{\mu} , \qquad g_{22} = -r^2 , \qquad g_{33} = -\, r^2 \sin^2\theta . \tag{21.3}$$

It follows at once from (11.8) that the contravariant tensor $g^{\mu\nu}$ is also diagonal and has the components:

$$g^{00} = e^{-\nu} ; \quad g^{11} = -\, e^{-\mu} , \quad g^{22} = \frac{-1}{r^2} , \quad g^{33} = \frac{-1}{r^2 \sin^2\theta} . \tag{21.4}$$

The next step is the calculation of the Christoffel symbols. As an example we calculate Γ^1_{00}. We find from (12.9):

$$\Gamma^1_{00} = \tfrac{1}{2} g^{11} (g_{10,0} + g_{01,0} - g_{00,1}) .$$

The first two terms vanish because of (21.3). Introducing the values (21.3) and (21.4) we find:

$$\Gamma^1_{00} = \tfrac{1}{2} e^{-\mu+\nu}\, \nu' , \qquad \nu' \equiv \frac{d\nu}{dr} .$$

The final result of all calculations of this type is the following:

$$\left.\begin{aligned}
&\Gamma^1_{11} = \tfrac{1}{2}\mu', && \Gamma^1_{00} = \tfrac{1}{2}e^{\nu-\mu}\,\nu', && \Gamma^2_{33} = -\sin\theta\cos\theta,\\
&\Gamma^1_{22} = -\,re^{-\mu}, && \Gamma^2_{12} = \frac{1}{r}, && \Gamma^3_{23} = \frac{\cos\theta}{\sin\theta},\\
&\Gamma^1_{33} = -\,re^{-\mu}\sin^2\theta, && \Gamma^3_{13} = \frac{1}{r}, && \Gamma^0_{10} = \tfrac{1}{2}\nu'.
\end{aligned}\right\} \tag{21.5}$$

The Christoffel symbols which do not appear in (21.5) are equal to zero.

We now have to calculate the Ricci tensor $R_{\mu\nu}$. We calculate as an example the component R_{00}. We find from (19.8):

$$R_{00} = -\,\Gamma^\alpha_{00,\,\alpha} + \Gamma^\alpha_{0\alpha,\,0} - \Gamma^\beta_{00}\Gamma^\alpha_{\beta\alpha} + \Gamma^\beta_{0\alpha}\Gamma^\alpha_{\beta 0}\,.$$

The second term vanishes because of (21.1). Remembering that the non-vanishing $\Gamma^\lambda_{\mu\nu}$ are those given in (21.5) we find:

$$R_{00} = -\,\Gamma^1_{00,\,1} - \Gamma^1_{00}\left(\Gamma^0_{10} + \Gamma^1_{11} + \Gamma^2_{12} + \Gamma^3_{13}\right) + 2\Gamma^1_{00}\Gamma^0_{10}\,.$$

Introducing the values (21.5) of the $\Gamma^\lambda_{\mu\nu}$ we find:

$$R_{00} = -\,\tfrac{1}{2}\left(e^{\nu-\mu}\nu'\right)' - \frac{\nu'}{2}\,e^{\nu-\mu}\left(\frac{\nu'+\mu'}{2} + \frac{2}{r}\right) + \frac{\nu'}{2}\,e^{\nu-\mu}\,\nu'\,.$$

The final result of all similar calculations is that the non-diagonal components $R_{\mu\nu}$ vanish identically. For the diagonal $R_{\mu\nu}$ we find:

$$\left.\begin{aligned}
&R_{00} = e^{\nu-\mu}\left\{-\frac{\nu''}{2} - \frac{\nu'}{r} + \frac{\nu'}{4}\left(\mu' - \nu'\right)\right\},\\
&R_{11} = \frac{\nu''}{2} - \frac{\mu'}{r} + \frac{\nu'}{4}\left(\nu' - \mu'\right),\\
&R_{22} = \frac{1}{\sin^2\theta}\,R_{33} = e^{-\mu}\left\{1 - e^{\mu} + \frac{r}{2}\left(\nu' - \mu'\right)\right\}.
\end{aligned}\right\} \tag{21.6}$$

We shall determine here the gravitational field outside of a central, spherically symmetric body. We therefore have to use the field Equations (19.4) for empty space. From the values (21.6) of $R_{\mu\nu}$ we get the following three equations:

$$\left.\begin{aligned}
&\frac{\nu''}{2} + \frac{\nu'}{r} - \frac{\nu'}{4}\left(\mu' - \nu'\right) = 0,\\
&\frac{\nu''}{2} - \frac{\mu'}{r} - \frac{\nu'}{4}\left(\mu' - \nu'\right) = 0,\\
&1 - e^{\mu} + \frac{r}{2}\left(\nu' - \mu'\right) = 0.
\end{aligned}\right\} \tag{21.7}$$

In (21.7) we have 3 equations for the 2 unknown functions μ and ν. There is, however, no overdetermination, because these equations are not independent. Indeed, we have the identity (14.14):

$$(R^\sigma_\varrho - \tfrac{1}{2}\delta^\sigma_\varrho R)_{;\,\sigma} = 0\,. \tag{21.8'}$$

This identity has in the general case 4 components. It is easy to verify that in the present case the components $\varrho = 0$, 2, 3 are satisfied in a trivial way. Only the component $\varrho = 1$ is not trivial and has the form:

$$(R^1_1 - R^0_0 - 2R^2_2)' + (R^1_1 - R^0_0)\,\nu' + \frac{4}{r}(R^1_1 - R^2_2) = 0 \tag{21.8}$$

giving one identical relation between the left-hand sides of the 3 equations (21.7).

It is very easy to find the general solution of (21.7), because these equations are total differential equations of a very simple form. Subtracting the second from the first of Equations (21.7) we find:

$$\mu' + \nu' = 0\,. \tag{21.9}$$

Therefore

$$\mu + \nu = \lambda\,, \qquad \lambda = \text{const.} \tag{21.10}$$

The last Equation (21.7) takes now the form:

$$1 - e^\mu - r\mu' = 0 \quad \Leftrightarrow \quad (re^{-\mu})' = 1\,. \tag{21.11}$$

The solution of this equation is:

$$e^{-\mu} = 1 - \frac{2m}{r}\,, \qquad m = \text{const}\,. \tag{21.12}$$

The constant of integration has been written in a form which will be convenient later. Now we get from (21.10):

$$e^\nu = e^{\lambda - \mu} = e^\lambda \left(1 - \frac{2m}{r}\right). \tag{21.13'}$$

The constant λ can be put equal to zero by a change of the time coordinate. Indeed, the first term of the right-hand side of (20.15) can be written as follows:

$$e^\nu\,\mathrm{d}t^2 = \left(1 - \frac{2m}{r}\right)\{\mathrm{d}\,(e^{\lambda/2}\,t)\}^2\,.$$

Writing again t instead of $e^{\lambda/2}t$ we shall have:

$$e^\nu = e^{-\mu} = 1 - \frac{2m}{r}\,. \tag{21.13}$$

This is the *Schwarzschild solution.* The Schwarzschild metric contains just one integration constant, the ds^2 being of the form:

$$ds^2 = \left(1 - \frac{2m}{r}\right) dt^2 - \left(1 - \frac{2m}{r}\right)^{-1} dr^2 - r^2 (d\theta^2 + \sin^2\theta \, d\varphi^2). \quad (21.14)$$

The physical meaning of the constant m will be determined in Section 22.

We repeat that the metric (21.14) describes the *exterior* gravitational field, outside of a central, spherically symmetric body. Note that the detailed distribution of matter in the interior of the body does not influence the exterior field, a feature which was present in the case of spherical symmetry also in the Newtonian theory of gravitation. Before discussing in detail the field determined by Equation (21.14) we shall examine a more general problem in order to establish one important result.

We shall consider a time-dependent spherically symmetric gravitational field, again satisfying the empty-space field equations. This may be for example the field surrounding a spherically symmetric star which is pulsating radially (which is compatible with the spherical symmetry). We shall have again the metric (20.15) with the two functions μ and ν depending now on t and r. In this case one verifies without difficulty that the Christoffel symbols given in (21.5) remain unchanged, the meaning of the prime being now $(..)' = (\partial/\partial r)\,(..)$. There are 3 additional Christoffel symbols which do not vanish identically. They are the following:

$$\Gamma^0_{00} = \tfrac{1}{2}\dot{\nu}, \qquad \Gamma^1_{10} = \tfrac{1}{2}\dot{\mu}, \qquad \Gamma^0_{11} = \tfrac{1}{2}e^{\mu-\nu}\dot{\mu}; \qquad (..)^{\cdot} = \frac{\partial}{\partial t}(..). \quad (21.15)$$

The components R_{22} and R_{33} of the Ricci tensor remain unchanged. The components R_{00} and R_{11} do get some additional terms:

$$\left.\begin{aligned} R_{00} &= \cdots + \tfrac{1}{2}(\ddot{\mu} + \tfrac{1}{2}\dot{\mu}^2 - \tfrac{1}{2}\dot{\mu}\dot{\nu}), \\ R_{11} &= \cdots - \tfrac{1}{2}e^{\mu-\nu}(\ddot{\mu} + \tfrac{1}{2}\dot{\mu}^2 - \tfrac{1}{2}\dot{\mu}\dot{\nu}), \end{aligned}\right\} \quad (21.16)$$

the omitted terms being those given in (21.6). Moreover, there is now one non-diagonal component $R_{\mu\nu}$ which does not vanish identically:

$$R_{01} = -\frac{1}{r}\dot{\mu}. \quad (21.17)$$

From the field equation

$$R_{01} = 0 \quad \Leftrightarrow \quad \dot{\mu} = 0 \quad (21.18)$$

we conclude at once that μ will again depend only on r. All extra terms appearing in (21.16) are now equal to zero and consequently we have again the Equations (21.7), the only difference being that ν is a function of t and r. The solution of these equations is found again in the same way as before. From the combination (21.9) of the two first Equations (21.7) we now derive the relation:

$$\mu + \nu = \lambda(t) \quad \Leftrightarrow \quad e^{\nu} = e^{\lambda(t)}e^{-\mu}. \quad (21.19)$$

The last Equation (21.7) gives again

$$e^{-\mu} = 1 - \frac{2m}{r} \tag{21.20}$$

with $\dot{m}=0$ because of (21.18). That is we have again $m=\text{const}$ as in the static case. Therefore the expression for $\mathrm{d}s^2$ is:

$$\mathrm{d}s^2 = e^{\lambda(t)}\left(1 - \frac{2m}{r}\right)\mathrm{d}t^2 - \left(1 - \frac{2m}{r}\right)^{-1}\mathrm{d}r^2 - r^2\left(\mathrm{d}\theta^2 + \sin^2\theta\,\mathrm{d}\varphi^2\right). \tag{21.21}$$

This expression differs from (21.14) by the time-dependent factor e^{λ} of the first term. This factor can be again made equal to 1, this time by a change $t \to \tilde{t} = f(t)$ of the time coordinate. Indeed, it is sufficient to use the $\tilde{t}$ given by

$$\tilde{t} = \int \exp\left[\tfrac{1}{2}\lambda(t)\right]\mathrm{d}t$$

in order to obtain exactly the Schwarzschild metric (21.14). Thus we have the following important result, which constitutes the *theorem of Birkhoff: A spherically symmetric solution of the vacuum field Equations* (19.4) *is necessarily static*. In the example considered earlier, a star pulsating radially, so that the spherical symmetry is maintained constantly, has the same external fields as a star (of the same mass) at rest.

We may recall that a similar result is valid in the Maxwell theory of the electromagnetic field in special relativity: A spherically symmetric solution of the vacuum Maxwell equations is necessarily static. The deeper physical meaning of this result will become clear later, after the discussion of the radiation problem (Section 48).

22. Geodesics in the Schwarzschild Space

We have seen in Section 18 that test particles move on geodesic lines. Therefore the study of geodesics is important, not only from a purely geometrical but also from the physical point of view.

The geodesic equation is given by (13.11):

$$\frac{\mathrm{d}^2 x^{\lambda}}{\mathrm{d}s^2} + \Gamma^{\lambda}_{\mu\nu}\frac{\mathrm{d}x^{\mu}}{\mathrm{d}s}\frac{\mathrm{d}x^{\nu}}{\mathrm{d}s} = 0. \tag{22.1}$$

As we mentioned in Section 13, a geodesic is determined when we give a point P on it and the direction of its tangent at P. In the present case, in which we consider only the space outside the central body, the point P and the direction of the tangent at P will define, together with the centre of symmetry O, a plane of the 3-dimensional space which is evidently a plane of symmetry of this space. It follows then by symmetry that the geodesic will lie entirely on this plane. This result will be derived also from the geodesic equation when we shall have it written in detail for the Schwarzschild metric.

Taking into account the values (21.5) for the non-vanishing Christoffel symbols we find the following components of the geodesic equation, the dot meaning now derivative with respect to s, $(.)^{\cdot}=(\mathrm{d}/\mathrm{d}s)\,(.)$:

$$\left.\begin{aligned}
&\ddot{r}-\tfrac{1}{2}v'\dot{r}^2-re^{v}\dot{\theta}^2-re^{v}\sin^2\theta\,\dot{\varphi}^2+\tfrac{1}{2}e^{2v}v'\dot{t}^2=0\,,\\
&\ddot{\theta}+\frac{2}{r}\dot{r}\dot{\theta}-\sin\theta\cos\theta\,\dot{\varphi}^2=0\,,\\
&\ddot{\varphi}+\frac{2}{r}\dot{r}\dot{\varphi}+\frac{2\cos\theta}{\sin\theta}\dot{\theta}\dot{\varphi}=0\,,\\
&\ddot{t}+v'\dot{r}\dot{t}=0\,.
\end{aligned}\right\}\qquad(22.2)$$

Let us consider a geodesic passing through a point P situated on the equatorial plane $\theta=\pi/2$ and having a tangent at P situated also in this plane:

$$\dot{\theta}=0\quad\text{at}\quad P\,.$$

We then derive from the second of Equations (22.2) differentiated repeatedly with respect to s:

$$\ddot{\theta}=\dddot{\theta}=\cdots=0\quad\text{at}\quad P\,.$$

This means that we shall have on this geodesic

$$\dot{\theta}=0\quad\Leftrightarrow\quad\theta=\frac{\pi}{2}\,.$$

That is, the geodesic lies entirely in the plane defined by P, the tangent at P and the centre of symmetry, as we concluded earlier by a reasoning based on the spherical symmetry of the space.

Since the symmetry planes are equivalent to each other, it will be sufficient to discuss the geodesics lying on one of these planes. It will be advantageous to choose the equatorial plane $\theta=\pi/2$, as this simplifies the Equations (22.2). The second of these equations is then satisfied automatically. Before discussing the remaining 3 Equations (22.2) we shall derive a general property of the geodesic equation, which will permit an additional simplification of the problem.

The 4 components of the geodesic equation are not independent. Indeed, there is a general first integral of these equations, which can be derived as follows. Using the relations

$$\begin{aligned}
&g_{\lambda\mu}\dot{x}^{\lambda}\ddot{x}^{\mu}=\tfrac{1}{2}g_{\lambda\mu}(\dot{x}^{\lambda}\dot{x}^{\mu})^{\cdot}\,,\\
&g_{\lambda\mu}\dot{x}^{\lambda}\Gamma^{\mu}_{\varrho\sigma}\dot{x}^{\varrho}\dot{x}^{\sigma}=\tfrac{1}{2}g_{\lambda\varrho,\sigma}\dot{x}^{\lambda}\dot{x}^{\varrho}\dot{x}^{\sigma}=\tfrac{1}{2}(g_{\lambda\varrho})^{\cdot}\,\dot{x}^{\lambda}\dot{x}^{\varrho}
\end{aligned}$$

we find, if we multiply (22.1) by $g_{\lambda\mu}\dot{x}^{\mu}$:

$$(g_{\lambda\mu}\dot{x}^{\lambda}\dot{x}^{\mu})^{\cdot}=0\,.$$

Consequently we have the following first integral:

$$g_{\lambda\mu}\dot{x}^{\lambda}\dot{x}^{\mu}=\text{const.}$$

Actually this relation is already known in the more concrete form (13.3):

$$g_{\lambda\mu}\dot{x}^\lambda \dot{x}^\mu = 1 . \tag{22.3}$$

Consequently we can replace any one of the 4 components of (22.1) by (22.3).

In the present case it will be advantageous to replace the first of the Equations (22.2) by (22.3). With the metric (21.14) the equation (22.3) takes the form:

$$e^\nu \dot{t}^2 - e^{-\nu}\dot{r}^2 - r^2\dot{\varphi}^2 = 1 . \tag{22.4}$$

Since the second Equation (22.2) is already satisfied, we have to consider besides (22.4) only the two last Equations (22.2) simplified by $\dot{\theta}=0$.

One can verify at once that these two equations have the following first integrals:

$$r^2\dot{\varphi} = \text{const} \equiv a , \tag{22.5}$$

$$e^\nu \dot{t} = \text{const} \equiv b . \tag{22.6}$$

These two integrals can be written in a much more intuitive form if we introduce the 4-dimensional *velocity* u^μ, according to (13.2),

$$u^\mu = \dot{x}^\mu ,$$

of a test particle which might be moving on this geodesic. Let us calculate the covariant components u_μ, using the diagonal metric (21.14) of the Schwarzschild solution. We find:

$$u_0 = g_{00}u^0 = e^\nu \dot{t} , \qquad u_1 = e^\mu \dot{r} , \qquad u_2 = - r^2 \dot{\theta}, \qquad u_3 = -r^2 \sin^2\theta\dot{\varphi} . \tag{22.7}$$

Remembering that in the present case $\theta=\pi/2$ we see at once that the Equations (22.5) and (22.6) can be written as follows:

$$u_3 = \text{const} , \tag{22.8}$$

$$u_0 = \text{const} . \tag{22.9}$$

Equation (22.9) is of a much more general validity, being true for any time-independent metric. This statement can be proved most easily if we use the form (13.23) of the geodesic equation:

$$\dot{u}_\mu = \tfrac{1}{2} g_{\alpha\beta,\mu} u^\alpha u^\beta . \tag{22.10}$$

A time-independent gravitational field is characterized by the relation (16.20). Therefore we shall have in this case the relation:

$$\dot{u}_0 = 0 ,$$

leading to the first integral (22.9). This first integral of the geodesic equation is directly related to the conservation of the energy of a test particle moving in a time-independent field: As we shall see in Section 42, the energy of the particle is given by $m_0 u_0$, m_0 being the (constant) rest-mass of the particle.

Equation (22.5) is essentially the conservation law for the angular momentum of the particle, the angular momentum being equal to

$$m_0 r^2 \dot{\varphi} = m_0 a\,. \tag{22.11}$$

Such a conservation law is valid more generally for any gravitational field which has an axis of symmetry. This can be seen again from the form (22.10) of the geodesic equation. If we use a coordinate system which has as coordinate x^3 the angle around the axis of symmetry and if the coordinate system is adapted to the symmetry, we shall have for the metric a relation of the type (16.20):

$$g_{\mu\nu,\,3} = 0\,. \tag{22.12}$$

Equation (22.10) leads then directly to the conservation law (22.8). These remarks illustrate the important fact that to any Killing vector of the space corresponds, because of Equation (22.10), a first integral of the geodesic equation, which is equivalent to a conservation law for the geodesic motion of a test particle.

Combining the Equations (22.5) and (22.6) with Equation (22.4) we can eliminate the proper time and so derive the equation of the 3-dimensional trajectory of the test particle. Since $\dot{r} = (\mathrm{d}r/\mathrm{d}\varphi)\,\dot{\varphi}$, we find:

$$e^{-\nu} b^2 - \left\{\frac{e^{-\nu}}{r^2}\left(\frac{\mathrm{d}r}{\mathrm{d}\varphi}\right)^2 + 1\right\}\frac{a^2}{r^2} = 1\,.$$

This equation can be written also in the form:

$$\left\{\frac{\mathrm{d}}{\mathrm{d}\varphi}\left(\frac{1}{r}\right)\right\}^2 + \frac{e^{\nu}}{r^2} = \frac{b^2 - e^{\nu}}{a^2}\,,$$

and if we introduce the expression (21.13) for e^{ν}:

$$\left\{\frac{\mathrm{d}}{\mathrm{d}\varphi}\left(\frac{1}{r}\right)\right\}^2 + \frac{1}{r^2} = \frac{b^2 - 1}{a^2} + \frac{2m}{a^2 r} + \frac{2m}{r^3}\,. \tag{22.13}$$

For the comparison with the corresponding equation of the Newtonian theory it will be preferable to differentiate this equation with respect to φ. The result is:

$$\frac{\mathrm{d}^2}{\mathrm{d}\varphi^2}\left(\frac{1}{r}\right) + \frac{1}{r} = \frac{m}{a^2} + \frac{3m}{r^2}\,. \tag{22.14}$$

The Newtonian trajectory is determined by eliminating the time t from the two equations expressing the conservation of energy and of angular momentum. If m_0 is the mass of the test particle and M the mass of the central body, these conservation laws are expressed by the equations:

$$\tfrac{1}{2} m_0 \left\{\left(\frac{\mathrm{d}r}{\mathrm{d}t}\right)^2 + r^2\left(\frac{\mathrm{d}\varphi}{\mathrm{d}t}\right)^2\right\} - \frac{G m_0 M}{r} = \text{const}\,; \tag{22.15}$$

$$r^2\frac{\mathrm{d}\varphi}{\mathrm{d}t} = \text{const} \equiv ac\,. \tag{22.16}$$

After eliminating the variable t from these two equations and differentiating with respect to φ we find:

$$\frac{d^2}{d\varphi^2}\left(\frac{1}{r}\right)+\frac{1}{r}=\frac{GM}{a^2c^2}. \tag{22.17}$$

Comparing (22.14) and (22.17) we see that these equations will differ only by the last term of (22.14) if we put:

$$m=\frac{GM}{c^2}. \tag{22.18}$$

The last term in (22.14) will then be the *relativistic correction* to the Newtonian equation (22.17). The relation (22.18) determines the physical meaning of the integration constant m contained in the Schwarzschild solution. Remembering that M is the mass of the central body, we may call the quantity m the mass of the body in relativistic units. As one sees from (22.13), m has the dimensions of a length. For this reason m is usually called the *gravitational radius* of the central body.

As a numerical example let us consider the case of the Sun. The mass of the Sun is $M \approx 2 \times 10^{33}$ g. With the value $G \approx 6.7 \times 10^{-8}$ dyn cm^2 g^{-2} of the gravitational constant we find:

$$m_{\text{Sun}} \approx 1.5 \text{ km}. \tag{22.19}$$

For the earth we find in the same way $m_{\text{Earth}} \approx 0.5$ cm.

The preceding equations will be used in Section 23 to study the planetary motions in general relativity. Indeed, the masses of the planets are much smaller than the mass of the Sun and consequently the planets can be considered as test particles. Here we shall make an estimate of the order of magnitude of the relativistic correction in the case of the planetary motions. First we remark that for trajectories which do not differ much from circles, as are the planetary trajectories, the second and third terms of Equation (22.14) are of the same order of magnitude. It follows that the order of magnitude of the relativistic correction relative to the other terms of Equation (22.14) is:

$$\frac{3m}{r^2}:\frac{1}{r}=\frac{3m}{r}\sim\frac{m}{r}.$$

The maximum value of m/r corresponds to the planet which is nearest to the Sun, i.e. to Mercury. The distance from Mercury to the Sun is $r \approx 5 \times 10^{12}$ cm. With the value (22.19) of $m = m_{\text{Sun}}$ we find for Mercury:

$$\frac{m}{r} \approx 10^{-7}.$$

The values of m/r for the other planets are smaller still. We thus have proved that the relativistic correction of the newtonian trajectories of the planets is very small indeed. The next step will be to determine quantitatively the consequences following from this correction, in order to make possible a comparison with the observations.

23. Advance of the Perihelion of a Planet

We consider first the Equation (22.17) of the Newtonian orbit of the planet. The general solution of this equation is:

$$\frac{1}{r} = \frac{m}{a^2}\{1 + \varepsilon\cos(\varphi + \varphi_0)\}, \tag{23.1}$$

containing two integration constants ε and φ_0. We simplify this solution by putting $\varphi_0 = 0$, this being equivalent to a change of the point corresponding to $\varphi = 0$. The constant ε is the eccentricity of the orbit which is an ellipse if $\varepsilon < 1$. Actually, the orbits of the planets are nearly circles, i.e. we have $\varepsilon \ll 1$. This remark allows the following simplified treatment of the relativistically corrected Equation (22.14).

In the last term of Equation (22.14), which is the small relativistic correction, we replace $1/r$ by the Newtonian value (23.1) and we omit the terms containing powers of ε higher than the first. Thus the relativistic equation takes the form:

$$\frac{d^2}{d\varphi^2}\left(\frac{1}{r}\right) + \frac{1}{r} = \frac{m}{a^2} + \frac{3m^3}{a^4}(1 + 2\varepsilon\cos\varphi). \tag{23.2}$$

The constant term $3m^3/a^4$ is a small correction added to m/a^2, therefore meaning a very small change of the length of the major axis of the ellipse. This change is not interesting, as we cannot determine lengths with accuracy. We can therefore omit this term and write instead of (23.2) the equation:

$$\frac{d^2}{d\varphi^2}\left(\frac{1}{r}\right) + \frac{1}{r} = \frac{m}{a^2} + \frac{6m^3}{a^4}\varepsilon\cos\varphi. \tag{23.3}$$

Equation (23.3) is a linear differential equation for $1/r$. The solution of this equation is a superposition of the solution (23.1) of the Newtonian equation (22.17) and of some special solution of (23.3) when the term m/a^2 has been suppressed. One can verify at once that such a special solution is the following:

$$\frac{1}{r} = \frac{3\varepsilon m^3}{a^4}\varphi\sin\varphi.$$

Therefore the solution of (23.3) is:

$$\frac{1}{r} = \frac{m}{a^2}\left\{1 + \varepsilon\cos\varphi + \frac{3\varepsilon m^2}{a^2}\varphi\sin\varphi\right\}. \tag{23.4}$$

The form (23.4) of the solution is not convenient for its geometric interpretation. Remembering from (22.14) that m/a^2 is of the order of $1/r$ and therefore m^2/a^2 is of the order of the small quantity m/r we rewrite Equation (23.4) as follows:

$$\frac{1}{r} = \frac{m}{a^2}\left\{1 + \varepsilon\cos\left(\varphi\left[1 - \frac{3m^2}{a^2}\right]\right)\right\} + \cdots, \tag{23.5}$$

the omitted terms being at least of the order $(m/r)^2$.

The smallest value of the distance r corresponds to the values of the angle φ satisfying the relation

$$\cos\left\{\varphi\left(1-\frac{3m^2}{a^2}\right)\right\}=1.$$

These values are:

$$\varphi=0,\quad \frac{2\pi}{1-\dfrac{3m^2}{a^2}},\quad 2\frac{2\pi}{1-\dfrac{3m^2}{a^2}},\ \ldots.$$

Since $m^2/a^2 \ll 1$, we can write:

$$\frac{2\pi}{1-\dfrac{3m^2}{a^2}}=2\pi\left(1+\frac{3m^2}{a^2}+\cdots\right)=2\pi+\frac{6\pi m^2}{a^2}+\cdots.$$

Thus we see that in the relativistic orbit the perihelion is no longer a fixed point, as it was in the newtonian elliptic orbit, but that it moves in the direction of the motion of the planet, advancing by the angle

$$\delta\varphi=\frac{6\pi m^2}{a^2} \tag{23.6}$$

with each complete revolution around the Sun. If we introduce in (23.6) the major semi-axis r_0 of the unperturbed ellipse, which is found from (23.1) to be

$$r_0=\frac{a^2}{m}, \tag{23.7}$$

we find:

$$\delta\varphi=\frac{6\pi m}{r_0}. \tag{23.8}$$

We only mention here without entering into the calculations that, if one includes in (23.3) the term which is quadratic in the eccentricity ε, one finds for the angle $\delta\varphi$ the formula:

$$\delta\varphi=\frac{6\pi m}{r_0(1-\varepsilon^2)}. \tag{23.9}$$

For nearly circular orbits, as are the orbits of the planets, we have $\varepsilon \ll 1$ and consequently the difference between (23.8) and (23.9) is unimportant.

If we introduce in (23.8) the value (22.19) of m for the Sun and the distance r_0 of the planet Mercury from the Sun, we find the extremely small angle

$$\delta\varphi \approx 0.1''.$$

This is, however, a secular effect which increases with the number of revolutions. If we take into account that the period of revolution of Mercury around the Sun is $T \approx 0.24$ yr, we find that the perihelion of Mercury will advance in one century by an angle of 43″. This is very nearly the residue in the motion of the perihelion of Mercury which remained unexplained in the newtonian theory of gravitation, when all perturbations caused by the other planets were taken into account. We see that there is a very good agreement between the theoretical prediction and the observations. For the other planets the relativistic correction is smaller because of the increasing distance from the Sun. Consequently the comparison with the observations is more difficult than for Mercury.

It has to be remarked that the Sun has a slow rotation and therefore cannot be exactly spherically symmetric. When the central body is not exactly spherical, there is already a certain rotation of the perihelion given by the Newtonian theory of gravitation. Earlier estimates of the oblateness of the Sun led to values of the corresponding Newtonian rotation of the perihelion of Mercury which were much smaller than 43″. Recently Dicke has attempted a direct determination of the oblateness of the Sun and he claims that this oblateness is so large that it should account for about 20% of the value of 43″. If Dicke's result is confirmed, the agreement of the theoretical prediction with the observations will be considerably less good.

24. The Deflection of Light Rays

The propagation of light can be considered, from the corpuscular point of view, as the motion of a large number of photons. These particles move with the speed of light because they have a vanishing rest-mass. A photon moving in the gravitational field of the Sun will of course be considered as a test particle. Therefore the motion of photons, i.e. the propagation of a light ray in the Schwarzschild field will be described by the null-geodesics of the Schwarzschild metric.

The results obtained in Section 22 from the general discussion of the geodesics in the Schwarzschild space can be used in this case too. We only have to remark that in the limit of a null geodesic we have $\mathrm{d}s \to 0$ and consequently, according to (22.5) and (22.6):

$$a, b \to \infty, \qquad \frac{a}{b} \text{ finite}.$$

With the value $a \to \infty$ the Equation (22.14) determining the 3-dimensional orbit of a test particle takes the simpler form:

$$\frac{\mathrm{d}^2}{\mathrm{d}\varphi^2}\left(\frac{1}{r}\right) + \frac{1}{r} = \frac{3m}{r^2}. \tag{24.1}$$

Putting $m=0$ in (24.1) we find the equation:

$$\frac{\mathrm{d}^2}{\mathrm{d}\varphi^2}\left(\frac{1}{r}\right) + \frac{1}{r} = 0. \tag{24.2}$$

The solution of this equation is:

$$\frac{1}{r} = \frac{1}{r_0} \cos(\varphi + \varphi_0). \tag{24.3}$$

The relation (24.3) determines a straight line in the plane described by the polar coordinates r and φ. The length r_0 is the distance of this line from the origin. The meaning of the second integration constant φ_0 is shown on Figure 5.

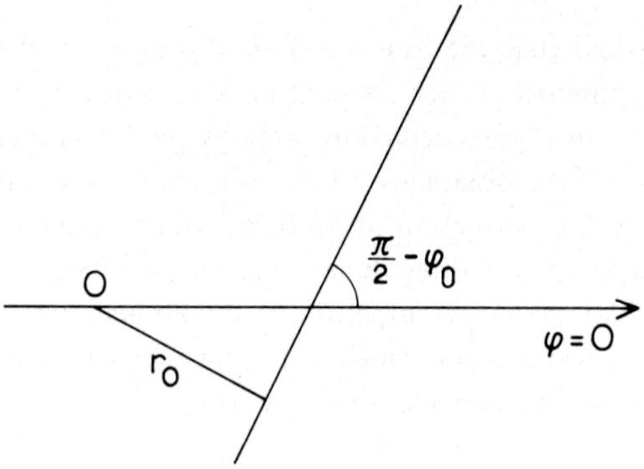

Fig. 5. The line represented by Equation (24.3).

Since the term on the right-hand side of Equation (24.1) is very small compared with the other terms, we shall obtain a good approximation if we replace in this term the quantity $1/r$ by the value (24.3). Putting for simplicity $\varphi_0 = 0$ we find the equation:

$$\frac{d^2}{d\varphi^2}\left(\frac{1}{r}\right) + \frac{1}{r} = \frac{3m}{r_0^2} \cos^2 \varphi. \tag{24.4}$$

This equation has the following special solution:

$$\frac{1}{r} = \frac{m}{r_0^2}(1 + \sin^2 \varphi). \tag{24.5}$$

The solution of (24.1) will be the sum of the solution (24.3) of the homogeneous equation and of the special solution (24.5):

$$\frac{1}{r} = \frac{1}{r_0} \cos \varphi + \frac{m}{r_0^2}(1 + \sin^2 \varphi). \tag{24.6}$$

On the line determined by (24.3) with $\varphi_0 = 0$ the points which are at a distance $r \to \infty$ correspond to the values of φ satisfying the condition:

$$\cos \varphi = 0 \quad \Leftrightarrow \quad \varphi = \pm \frac{\pi}{2}.$$

This agrees with the fact that this line is a straight line orthogonal to the line $\varphi = 0$.

Now, on the line determined by the Equation (24.6) we shall have $r \to \infty$ for $\varphi = \pm[(\pi/2) + \alpha]$, the angle α satisfying the relation:

$$-\frac{1}{r_0}\sin\alpha + \frac{m}{r_0^2}(1 + \cos^2\alpha) = 0. \tag{24.7}$$

Since the angle α is very small, we write:

$$\sin\alpha = \alpha + \cdots, \qquad \cos\alpha = 1 + \cdots$$

and we find then, omitting terms of higher order:

$$\alpha = \frac{2m}{r_0}. \tag{24.8}$$

The line determined by Equation (24.6) is the trajectory of a photon or of a light ray. Consequently the deflection of the light ray will be:

$$\delta = 2\alpha = \frac{4m}{r_0}. \tag{24.9}$$

A light ray coming towards us from a star situated nearly behind the Sun will have the maximum deflection if it passes just outside the Sun's surface. The radius of the Sun is $r_0 \approx 7 \times 10^{10}$ cm. With the value (22.19) of m we find that the maximum deflection will be

$$\delta \approx 1.75''. \tag{24.10}$$

It has to be remarked that a bending of the orbit of a photon, i.e. of a particle moving with the speed c, is predicted also by the Newtonian theory of gravitation. Of course, in this case the velocity c does not play any special role, as in special or in general relativity. The constant a entering in the Equation (22.16) has a value which can be derived at once if we consider the particle at its position of minimum distance r_0 from the origin:

$$ac = r_0^2 \frac{\mathrm{d}\varphi}{\mathrm{d}t} = r_0 c, \quad \text{i.e.} \quad a = r_0.$$

Thus the Newtonian equation of the orbit (22.17) will be:

$$\frac{\mathrm{d}^2}{\mathrm{d}\varphi^2}\left(\frac{1}{r}\right) + \frac{1}{r} = \frac{m}{r_0^2}. \tag{24.11}$$

The solution of this equation, which is relevant now, will describe a hyperbola differing very little from a straight line. This solution is:

$$\frac{1}{r} = \frac{m}{r_0^2} + \frac{1}{r_0}\cos\varphi.$$

From this equation one derives at once that the newtonian light deflection is equal to one half of the relativistic deflection (24.9):

$$\delta_{\text{Newt.}} = \tfrac{1}{2}\delta_{\text{rel}} = \frac{2m}{r_0}.$$

The preceding discussion of the light deflection has been based on the corpuscular aspect of light. The same problem can be treated also on the basis of the wave aspect. Remembering that light is an electromagnetic wave we see that for this discussion one has to start from the Einstein-Maxwell equations for the combined gravitational and electromagnetic field (Section 36). More exactly one has to discuss the propagation of a test electromagnetic wave in the gravitational field of the Sun (Schwarzschild field). We shall only mention that this problem has been discussed in detail. The result obtained in this way is identical, in the approximation of geometrical optics, with the result obtained here from the corpuscular point of view.

For the experimental verification of the deflection effect one has to start by observing photographically the starfield surrounding the Sun's disc during a total eclipse. The same starfield will be photographed again several months later, when it is sufficiently far from the Sun. The comparison of the two photographs allows one to determine the angle δ. The observations made until now lead to an average value $\delta \approx 2''$, which is about 20% larger than the theoretical prediction. In view of the extreme difficulties of these observations and also of the fact that it has not been possible until now to repeat them as often as needed, we may consider the agreement between theory and observation as qualitatively good. Of course, the situation has to be clarified in the future.

25. Red Shift of Spectral Lines

The problem which we shall examine now is the following. Let us consider two identical atoms, e.g. hydrogen atoms, one on the surface of the Sun and the other in a laboratory on the Earth. The first atom emits one of its characteristic spectral lines in the direction of the laboratory, so that we can observe the emitted radiation and determine its wavelength. What will be this wavelength compared with the wavelength of the same spectral line emitted by the second atom in the laboratory?

In order to find the solution of this problem we have to consider separately two different questions. The first question is the following. What can we derive from the principle of equivalence for these two atoms? We shall assume that both atoms are *falling freely*, i.e. that they are not acted upon by any non-gravitational forces, at the moment of emission. If we examine the first atom in a frame comoving with it, we shall find that the gravitational acceleration has been eliminated. A similar remark holds for the second atom. Consequently for an observer comoving with the first atom the behaviour of this atom will be identical with the behaviour of the second atom for a second observer comoving with it. In particular the two observers will find exactly the same frequency of the emitted spectral line. Since a comoving observer measures the *proper frequency*, we have the result: The spectral lines emitted by identical freely falling atoms have *the same proper frequencies*. Or, in the corpuscular description, the photons emitted by identical atoms will have, for observers comoving with each of the atoms, exactly the *same energy*.

Let us now consider an observer moving on the worldline L (Figure 6) and assume that at the point P the observer meets a particle having at that moment the energy-momentum vector p^μ. What is the energy of the particle as measured by this observer?

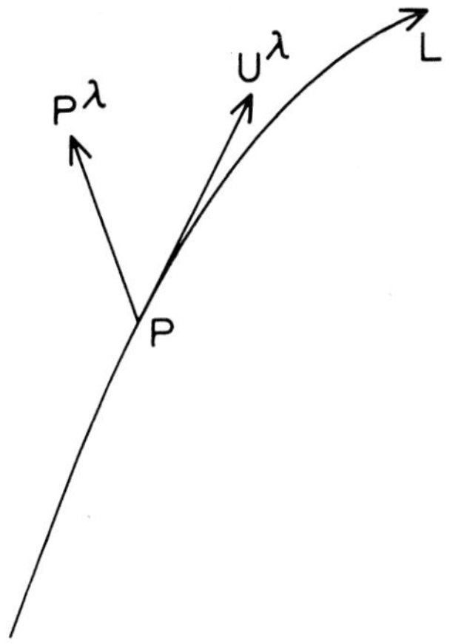

Fig. 6. An observer determines the energy of a particle.

The answer to this question will be found at once if we consider the transformation $x^\mu \to \tilde{x}^\mu$, the new coordinate system $\tilde{x}^\mu$ satisfying the following two conditions:

(1) The metric $\tilde{g}_{\mu\nu}$ has at the point P the Minkowski form (15.4).

(2) The coordinate system $\tilde{x}^\mu$ is comoving with the observer at P, i.e. the velocity $\tilde{U}^\mu$ of the observer at P is:

$$\tilde{U}^\mu = (1, 0, 0, 0).$$

If the vector p^μ is transformed in $\tilde{x}^\mu$ into

$$\tilde{p}^\mu = (\tilde{p}^0, \tilde{p}^1, \tilde{p}^2, \tilde{p}^3),$$

then the energy of the particle as measured by this observer will be

$$E = c\tilde{p}^0 = c\tilde{p}_0. \tag{25.1}$$

Now we verify at once the relation:

$$\tilde{p}_0 = \tilde{g}_{\mu\nu}\tilde{p}^\mu\tilde{U}^\nu = \tilde{p}_\nu\tilde{U}^\nu. \tag{25.2}$$

Since the right-hand side is a scalar, we can calculate it in any coordinate system. Therefore the energy of a particle having the energy-momentum vector p^μ for an observer having the velocity U^μ is:

$$E = cp_\mu U^\mu. \tag{25.3}$$

We now apply this formula to the two photons emitted at P and Q (Figure 7), the corresponding energy-momentum vectors of the photons being $p^\mu(P)$ and $p^\mu(Q)$. Let the velocities of the comoving observers, i.e. the velocities of the emitting atoms be $U^\mu(P)$ and $U^\mu(Q)$. Since each photon has the same energy for an observer comoving with the corresponding atom, we find from (25.3)

$$\frac{E}{c} = p_\mu(P)\, U^\mu(P) = p_\mu(Q)\, U^\mu(Q). \tag{25.4}$$

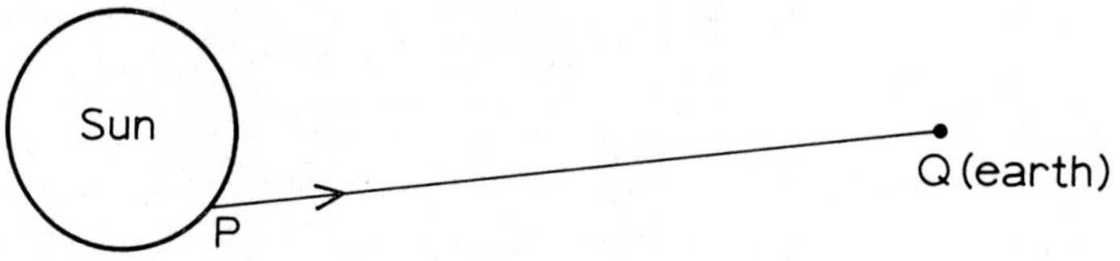

Fig. 7. Comparison of frequencies of photons emitted on the surface of the Sun and in the laboratory.

The photon emitted at P will move through the space as a test particle to arrive later at the point Q. The equation of motion of a particle having a non-vanishing rest-mass m_0 is given by (18.5) or (13.24):

$$u_{\mu;\nu} u^\nu = 0. \tag{25.5}$$

The rest-mass m_0 being constant, we can rewrite this equation using the energy-momentum vector p_μ of the particle,

$$p_\mu = m_0 u_\mu, \tag{25.6}$$

instead of u_μ. The result is:

$$p_{\mu;\nu} p^\nu = 0. \tag{25.7}$$

For a particle of zero rest-mass the unitary vector u_μ does not exist because of $ds=0$. However, the product $m_0 u_\mu$ exists and is the energy-momentum vector of the particle. Therefore the equation of motion (25.7) will be valid also for photons. This equation can be transformed to the following form:

$$p_{\mu,\nu} p^\nu = \tfrac{1}{2} g_{\lambda\nu,\mu} p^\lambda p^\nu, \tag{25.8}$$

which is the analogon of the form (13.23) of Equation (25.5).

We shall apply the preceding formulae to a case like the one shown on Figure 7. In order to simplify the problem we shall neglect the Doppler effect due to the relative motion of the two atoms by assuming that both atoms are, at the moment of the emission, at rest in the coordinate system in which the gravitational field of the Sun is described by the metric (21.14). (The velocity of the Earth around the Sun is appreciable, but it has a direction very nearly orthogonal to the line PQ.)

Since the two atoms are assumed to be at rest, their velocities $U^\mu(P)$ and $U^\mu(Q)$ will satisfy the relations:

$$U^i(P) = U^i(Q) = 0, \qquad i = 1, 2, 3. \tag{25.9}$$

Therefore from (25.4):

$$p_0(P)\, U^0(P) = p_0(Q)\, U^0(Q). \tag{25.10}$$

From the relation (13.3) we find:

$$g_{00}(P)\,\{U^0(P)\}^2 = g_{00}(Q)\,\{U^0(Q)\}^2 = 1. \tag{25.11}$$

Combining (25.10) and (25.11) we obtain the result:

$$p_0(P)/p_0(Q) = \{g_{00}(P)/g_{00}(Q)\}^{1/2}. \tag{25.12}$$

The photon emitted at P will move towards the point Q with its energy-momentum vector p_μ satisfying the equation of motion (25.8). Since the field is time-independent, we see at once from the component $\mu=0$ of (25.8) that the component p_0 remains constant and it will therefore have the same value when the photon arrives at Q. We write this result as follows:

$$p_0(P \to Q) = p_0(P). \tag{25.13}$$

Therefore from (25.1) and (25.12):

$$\frac{E(P \to Q)}{E(Q)} = \frac{p_0(P \to Q)}{p_0(Q)} = \left\{\frac{g_{00}(P)}{g_{00}(Q)}\right\}^{1/2}. \tag{25.14}$$

The two photons being now at the point Q and observed by the same observer, their energies will be, because of the Planck relation:

$$E = h\nu = hc/\lambda,$$

proportional to the corresponding frequencies or inversely proportional to their wavelengths. Therefore the corresponding wavelengths will satisfy the relation:

$$\frac{\lambda(P \to Q)}{\lambda(Q)} = \left\{\frac{g_{00}(Q)}{g_{00}(P)}\right\}^{1/2}. \tag{25.15}$$

Note that in the derivation of this formula we did not use the Schwarzschild metric explicitly: The result is valid *for any time-independent* gravitational field.

We now apply the formula (25.15) to the case shown on Figure 7. The gravitational field of the Sun is the Schwarzschild field (21.14). Therefore:

$$g_{00}(P) = 1 - \frac{2m}{r_S}, \qquad g_{00}(Q) = 1 - \frac{2m}{R}; \tag{25.16}$$

m is the quantity (22.19), r_S the radius of the Sun and R the distance Sun–Earth. Hence from (25.15):

$$\frac{\lambda(P \to Q)}{\lambda(Q)} = \left(1 - \frac{2m}{R}\right)^{1/2} \left(1 - \frac{2m}{r_S}\right)^{-1/2} \approx 1 + \frac{m}{r_S} - \frac{m}{R}.$$

The last term is nearly 100 times smaller than m/r_S and it can therefore be neglected. We see that the wavelength of the photon emitted on the Sun's surface is longer than that of an identical atom on the Earth: There is a *red-shift* $\delta\lambda$ of the spectral line:

$$\delta\lambda = \lambda(P \to Q) - \lambda(Q), \tag{25.17}$$

for which we get the following final formula:

$$\frac{\delta\lambda}{\lambda} = \frac{m}{r_S}. \tag{25.18}$$

Introducing in this formula the values $m \approx 1.5 \times 10^5$ cm and $r_S \approx 7 \times 10^{10}$ cm we find:

$$\frac{\delta\lambda}{\lambda} \approx 2 \times 10^{-6}. \tag{25.19}$$

We have to mention that the result (25.15) can be derived in a somewhat simpler manner by a calculation based on a wave picture of the spectral line, but only for the case of the sources P and Q at rest in a time-independent gravitational field. A qualitative reasoning based on the notion of the (Newtonian) gravitational potential energy leads to the same result. The method which has been used here has the advantage that it can be applied also to the case of a time-dependent gravitational field with the sources P and Q moving in any given manner.

The comparison of the theoretical prediction with the observations leads to results which are complicated and depend on the spectral line which has been selected for the observation. The observed red shift depends also on the position of the source P on the Sun's disc: It is smaller for light coming from the central part of the disc and increases towards the limb.

These complications can be explained qualitatively as the consequences of the following two phenomena:

(1) The Sun's photosphere as a whole is in equilibrium, the gravitational force being balanced by a corresponding pressure-gradient. Therefore an atom is falling freely only during the time-interval between two consecutive collisions with other atoms. The collisions cause a shift of the spectral lines known as the pressure effect. It is this effect which should explain the fact that the red-shift depends on the spectral line used in the observation. A quantitative comparison is not possible because the value of the pressure effect under the conditions prevailing in the Sun's photosphere cannot be determined.

(2) The general variation of the observed red shift from centre to limb can be explained as the consequence of a *turbulent motion* in the Sun's photosphere. The ascending elements are hotter than the descending ones with the consequence of a

small residual *violet* Doppler shift, which cancels a part of the relativistic red shift. This Doppler shift has its maximum value at the centre and vanishes on the limb. In general one assumes at present that the gravitational red shift is predicted correctly by Equation (25.18). The discrepancies between the theoretical prediction and the observations are then used as possible sources of information about phenomena taking place in the Sun's photosphere.

Since 1960 there is also a laboratory test of the gravitational red shift, using the gravitational field of the earth. This effect is much weaker, of the order $\delta\lambda/\lambda \approx 10^{-15}$. The observation is possible through the Mössbauer effect, permitting the emission of γ-rays of extremely narrow profile, which are then reabsorbed with a correspondingly sharp resonance. There is good agreement with the theoretical prediction.

Another experimental test of general relativity will be mentioned here without any calculations. An electromagnetic signal emitted from the earth is reflected, for example, on a planet and then received on the earth. The time interval from the moment of emission of the signal until its return has a slightly larger value according to general relativity than according to the Newtonian theory of gravitation, the difference being larger when the signal travels nearer to the Sun. These *time-delay* experiments have already led to preliminary results which are in a qualitative agreement with the theoretical prediction.

26. The Schwarzschild Sphere. The (Event) Horizon. Kruskal Coordinates

In the case of the Sun the gravitational field is described by the Schwarzschild metric (21.14) only for distances $r \geqslant r_S$, r_S being the radius of the Sun, $r_S \approx 7 \times 10^{10}$ cm. The gravitational radius of the Sun is according to (22.19) $m_S \approx 1.5 \times 10^5$ cm and consequently the components $g_{\mu\nu}$ will differ from the corresponding components of the Minkowski metric by terms of the order $2m_S/r_S \approx 10^{-5}$. We say in this case that the gravitational field is *weak*. In the case of a white dwarf this difference may be of the order 10^{-3} and for neutron stars the gravitational field is no longer weak. Moreover there exists in general relativity, as we shall see in Section 27, the possibility of *gravitational collapse* of a star, with its total mass contracting to a singularity. It is therefore interesting, when discussing the Schwarzschild metric, to assume it to be valid for arbitrarily small values of r.

The standard form (21.14) of the Schwarzschild metric shows an anomalous behaviour at the points of the *Schwarzschild sphere* of radius $r = 2m$. A first, elementary anomaly is that we have on this sphere

$$g_{00} = g^{11} = 0, \qquad g^{00} = g_{11} = \infty .$$

But the quantity $\det g_{\mu\nu}$ remains finite and, as we shall prove in detail later, there is no singularity of the metric at $r = 2m$.

A more serious anomaly is the following. One can verify at once using the Equations (22.2) that the parametric lines of the coordinate r, i.e. the lines on which the coordinates t, θ and φ have constant values, are geodesics. But these geodesics are space-like for $r > 2m$ and time-like for $r < 2m$. The tangent vector of a geodesic undergoes parallel transport along the geodesic and consequently it cannot change from a time-like to a space-like vector. It follows that the two regions $r > 2m$ and $r < 2m$ do not join smoothly on the surface $r = 2m$.

This can be seen in a more striking manner if we consider the radial null directions, on which $d\theta = d\varphi = 0$. We have then:

$$ds^2 = \left(1 - \frac{2m}{r}\right) dt^2 - \left(1 - \frac{2m}{r}\right)^{-1} dr^2 = 0 .$$

Consequently the radial null directions satisfy the relation:

$$\frac{dr}{dt} = \pm \left(1 - \frac{2m}{r}\right). \tag{26.1}$$

If we take into account the fact that the time-like directions are contained in the light-cone, we find that in the region $r > 2m$ the light cones have, in the plane (r, t), the orientation shown on Figure 8. The opening of the light cone, which is nearly equal to $\pi/4$ for $r \gg 2m$, decreases with r and tends to zero when $r \to 2m$. On the contrary, in the region $r < 2m$ the parametric lines of the coordinate t are space-like and consequently

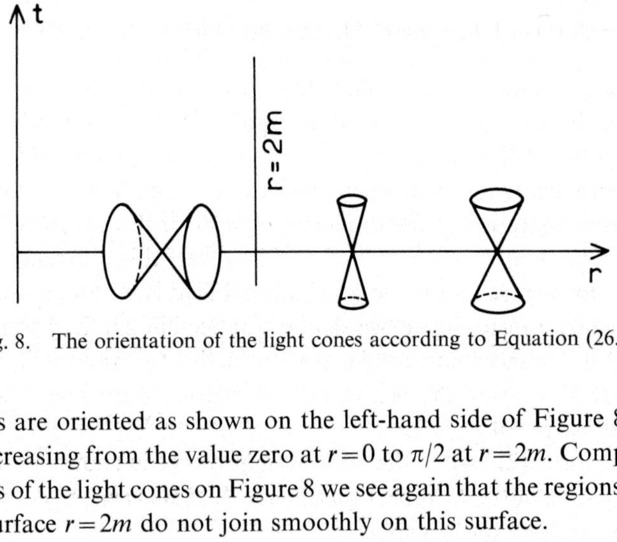

Fig. 8. The orientation of the light cones according to Equation (26.1).

the light cones are oriented as shown on the left-hand side of Figure 8, the opening of the cone increasing from the value zero at $r = 0$ to $\pi/2$ at $r = 2m$. Comparing the two different forms of the light cones on Figure 8 we see again that the regions on either side of the hypersurface $r = 2m$ do not join smoothly on this surface.

We shall obtain a better set of coordinates with the help of the transformation:

$$\left.\begin{aligned} t &= t' \pm 2m \lg\left(\frac{r'}{2m} - 1\right) \quad \text{for} \quad r > 2m , \\ t &= t' \pm 2m \lg\left(1 - \frac{r'}{2m}\right) \quad \text{for} \quad r < 2m ; \\ r &= r' , \qquad \theta = \theta' , \qquad \varphi = \varphi' . \end{aligned}\right\} \tag{26.2}$$

Introducing these relations in (21.14) we find:

$$\mathrm{d}s^2 = \left(1 - \frac{2m}{r}\right)\mathrm{d}t'^2 - \left(1 + \frac{2m}{r}\right)\mathrm{d}r^2 \pm \frac{4m}{r}\,\mathrm{d}t'\,\mathrm{d}r - r^2(\mathrm{d}\theta^2 + \sin^2\theta\,\mathrm{d}\varphi^2). \tag{26.3}$$

This is the Eddington-Finkelstein form of the Schwarzschild metric which we shall now discuss in some detail. Note that (26.3) gives in reality two different forms of the metric, corresponding to the sign + or − of the third term. We shall refer to them in the following as the Equations $(26.3)_+$ and $(26.3)_-$.

We first remark that the new form (26.3) of the metric is again time-independent, but it is now non-diagonal. One sees at once that the new metric is regular on the surface $r=2m$. As we shall show in the following, the two regions on either side of the surface $r=2m$ do now join smoothly on this surface.

A more detailed understanding of the situation will be obtained if we examine the general structure of the null cones of the metric (26.3) in the plane (t', r). We shall discuss first the case $(26.3)_-$. The radial null directions – θ and φ constant – will be determined by the equation:

$$\mathrm{d}s^2 = \left(1 - \frac{2m}{r}\right)\mathrm{d}t'^2 - \left(1 + \frac{2m}{r}\right)\mathrm{d}r^2 - \frac{4m}{r}\,\mathrm{d}t'\,\mathrm{d}r = 0\,.$$

This relation can be written in the form:

$$(\mathrm{d}t' + \mathrm{d}r)\left\{\left(1 - \frac{2m}{r}\right)\mathrm{d}t' - \left(1 + \frac{2m}{r}\right)\mathrm{d}r\right\} = 0\,.$$

It follows that the two null directions are:

$$\frac{\mathrm{d}r}{\mathrm{d}t'} = -1 \quad \text{and} \quad \frac{\mathrm{d}r}{\mathrm{d}t'} = \frac{r - 2m}{r + 2m}\,. \tag{26.4}$$

The lines which are tangent to one or the other of the directions (26.4) are the radial null-geodesics. We see at once that the first family of null geodesics has the very simple equation

$$t' + r = \text{const.} \tag{26.5}$$

These are the straight parallel lines shown on Figure 9. The second Equation (26.4) has a less simple integral. One sees, however, at once that the tangent to the null geodesics of this second family has the properties:

$$\frac{\mathrm{d}r}{\mathrm{d}t'} \to 1 \quad \text{for} \quad r \to \infty\,; \qquad \frac{\mathrm{d}r}{\mathrm{d}t'} \to 0 \quad \text{for} \quad r \to 2m\,;$$

$$\frac{\mathrm{d}r}{\mathrm{d}t'} \to -1 \quad \text{for} \quad r \to 0\,.$$

It follows from the second of these properties that the geodesics do not cross the surface $r=2m$ and will therefore be of the form shown on Figure 9.

It is easy to prove, starting from (26.3), that the non-radial null directions as well as the time-like directions have values $\mathrm{d}r/\mathrm{d}t'$ contained between the two values (26.4). Consequently the light cones will have on the plane (t', r) the form shown on Figure 9.

We now recall that physical particles move on time-like worldlines or on null-lines, i.e. on lines which lie inside or on the surface of the light cones. It follows then from Figure 9 first of all that no particle can cross the hypersurface $r=2m$ outwards. Moreover, any particle which is at some moment inside the surface $r=2m$ will necessarily move towards the line $r=0$, reaching it in finite coordinate time as well as proper time.

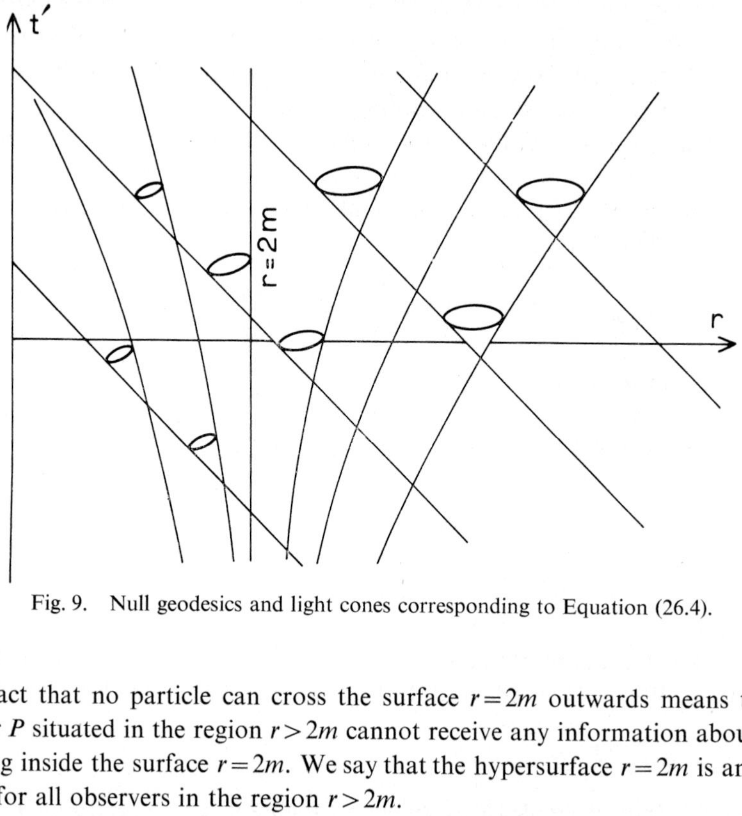

Fig. 9. Null geodesics and light cones corresponding to Equation (26.4).

The fact that no particle can cross the surface $r=2m$ outwards means that any observer P situated in the region $r>2m$ cannot receive any information about events occurring inside the surface $r=2m$. We say that the hypersurface $r=2m$ is an (event) *horizon* for all observers in the region $r>2m$.

The horizon is characterized by the following simple geometrical property: The normal to this hypersurface is a null vector. Indeed, the normal to the hypersurface defined by the equation

$$f(x^\mu) = r - 2m \equiv x^1 - 2m = 0$$

is the vector

$$n_\mu = f_{,\mu} = (0, 1, 0, 0)\,.$$

The contravariant form $g^{\mu\nu}$ of the metric (26.3) is determined from Equation (11.8) and has the components:

$$g^{00} = 1 + \frac{2m}{r}, \qquad g^{11} = -\left(1 - \frac{2m}{r}\right), \qquad g^{01} = \pm \frac{2m}{r}, \tag{26.6}$$

the remaining components being the same as in (21.4). Therefore we have $g^{11}=0$ at $r=2m$ and the vector n_μ is a null vector. Hypersurfaces with this property are called *characteristic surfaces*. They play an important role in the study of radiation and will be discussed in detail in Section 47. The horizon is a characteristic surface which is contained in a bounded region of the 3-dimensional space.

The case of the metric $(26.3)_+$ will be discussed very briefly as it can be derived from the metric $(26.3)_-$ by a time inversion. It follows that in this case we shall have instead of Figure 9 the new Figure 10, resulting from the preceding one by a reflexion with respect to the axis Or. We see from Figure 10 that now no particles can cross the horizon inward and that particles situated at some moment in the region $r<2m$ will necessarily move outwards and reach the horizon in finite proper time.

The coordinates used in (26.3) have, compared with those used in (21.14), the advantage that they describe the neighbourhood of the hypersurface $r=2m$ in a satisfactory way. However, the metric (26.3) has still a certain deficiency, as we shall explain now. Let us consider a radial null geodesic PQ (Figure 10) situated in the region $r>2m$ and proceeding towards decreasing values of r. This geodesic will reach the surface $r=2m$ asymptotically at a point Q_∞ for $t' \to \infty$.

One can verify easily that if one uses on this geodesic the time t' as a parameter, it will not be an affine parameter (see Section 8). We can determine an affine parameter

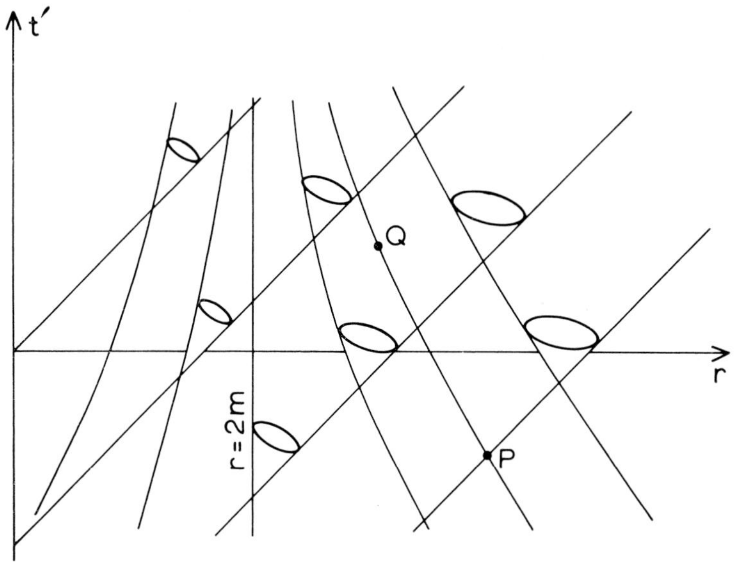

Fig. 10. Radial null geodesics and light cones of the metric $(26.3)_+$.

with the help of Equation (8.7). The result is extremely simple: The coordinate r is an affine parameter. It follows that the interval of affine parameter, which plays the role of the interval of proper time on the null geodesics, remains finite from the point P to Q_∞ and has the value $r_P - 2m$.

A similar result is derived for any time-like geodesic PQ' starting from a point P in $r > 2m$ and proceeding towards decreasing values of r: The geodesic reaches the surface $r = 2m$ asymptotically at a point Q'_∞ for $t' \to \infty$, the proper length of the arc PQ'_∞ remaining finite. To express these facts we say that the space described by these coordinates is *incomplete* in the directions defined by $r = 2m$ and $t' \to \infty$.

One can verify without difficulty that we have a similar situation in the metric $(26.3)_-$; this time the incompleteness is in the directions defined by $r = 2m$ and $t' \to -\infty$ (see Figure 9).

This deficiency is avoided in the Kruskal coordinates (w, v, θ, φ), which describe a geodesically complete space. The new coordinates are defined by the coordinate transformation:

$$\left.\begin{aligned} w &= \tfrac{1}{2} e^{r/4m} \left(\frac{r - 2m}{2m} e^{t'/4m} + e^{-t'/4m} \right), \\ v &= \tfrac{1}{2} e^{r/4m} \left(\frac{r - 2m}{2m} e^{t'/4m} - e^{-t'/4m} \right), \end{aligned}\right\} \tag{26.7}$$

t' and r being the coordinates used in $(26.3)_+$. We have the following simpler relations:

$$w + v = e^{(r+t')/4m} \frac{r - 2m}{2m}, \qquad w - v = e^{(r-t')/4m}. \tag{26.8}$$

The coordinates θ and φ remain unchanged.

From (26.8) we derive the relations

$$\frac{r - 2m}{2m} e^{r/2m} = w^2 - v^2, \qquad e^{t'/2m} = \frac{2m}{r - 2m} \frac{w + v}{w - v}, \tag{26.9}$$

which determine the inverse transformation $(w, v) \to (t', r)$.

Introducing the relations (26.9) in $(26.3)_+$ we find:

$$ds^2 = f^2(dv^2 - dw^2) - r^2(d\theta^2 + \sin^2\theta \, d\varphi^2); \tag{26.10}$$

$$f^2 = \frac{32m^3}{r} e^{-r/2m}. \tag{26.11}$$

The quantity r remaining in the factor f^2 in (26.10) is to be considered as the function of w and v defined by the first of Equations (26.9).

In these new coordinates the radial null geodesics have an extremely simple form. Indeed, they are defined by the condition

$$dw^2 - dv^2 = 0 \tag{26.12}$$

and consequently they have the equations

$$w \pm v = \text{const}\,, \tag{26.13}$$

being represented on the plane (w, v) by two sets of parallel lines (Figure 11).

Because of the simple form of the radial null geodesics these coordinates are very convenient for the qualitative discussion of several phenomena, e.g. the gravitational collapse (see Section 27). They have, however, the following strange feature: The transformation (26.7) maps the (t', r) plane on half of the (w, v) plane. It follows that when we use the whole of the (w, v) plane – this being necessary in order to obtain a complete space – there will be a *second space* (t', r). Whether this most unintuitive feature can have some physical significance is still an open question.

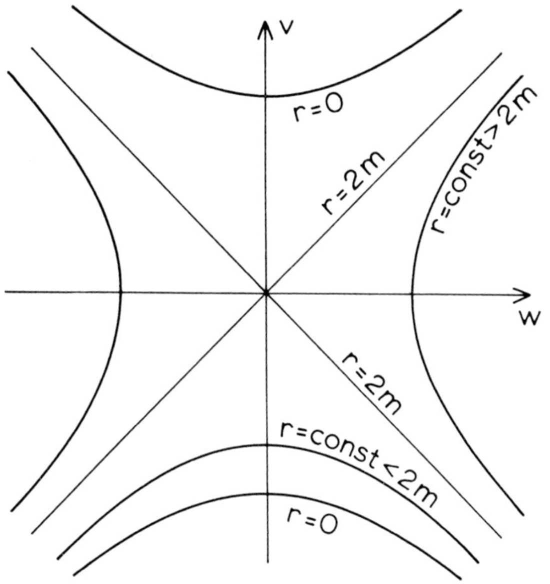

Fig. 11. The Kruskal diagram.

27. Gravitational Collapse. Black Holes

If the central body which is the source of the Schwarzschild field has a radius $a > 2m$, the forms (21.14) and (26.3) of the metric are equivalent, being related to each other by the transformation (26.2). The situation changes if $a < 2m$, i.e. when the Schwarzschild metric is valid down to values $r < 2m$. Indeed, in this case the form (21.14) cannot be used because of the difficulties on the surface $r = 2m$. Moreover, the two different forms contained in (26.3) describe essentially different situations, as it has already been shown by the discussion which led to the Figures 9 and 10.

The types of physical problems which are described by each of the two forms (26.3) of the metric will become clearer with the help of the following general remarks. Let us consider an (isolated) spherically symmetric star. Because of the radiation emitted by the star its temperature will finally start to decrease. Consequently the star will contract, i.e. the radius a of the star will be a decreasing function $a(t)$ of the time t. What will be the radius a_0 of the completely *cold* star?

In the Newtonian theory of gravitation we can have $a_0 \neq 0$ for any value of the total mass of the star, if the equation of state of the material of the star determines a pressure which increases sufficiently steeply with the density. In general relativity the fact that the stresses also are sources of the gravitational field leads to an entirely new situation. Indeed, the detailed discussion of this problem, into which we shall not enter here, has shown that there exists a positive a_0 only if the total mass M of the cold star is smaller than a certain limit M_0. For $M > M_0$ a positive a_0 does not exist and this for any possible equation of state. According to present evaluations the limit M_0 is a few times larger than the mass of the Sun.

Thus in general relativity a star of mass $M > M_0$ will necessarily contract until all matter contained in the star arrives at the center of symmetry. This is the *gravitational collapse*. We see at once from Figure 9 that this phenomenon can be described only within the form $(26.3)_-$ of the metric. The curve $r = a(t)$ represents the motion of the material situated on the surface of the star and must therefore be time-like everywhere. This curve divides the (t', r) plane in two. In the upper part of the plane we have the Schwarzschild solution valid everyhere, down to the value $r = 0$. In the lower part the gravitational field will be given by the *interior solution*, corresponding to the actual distribution of matter inside the star. It follows that the incompleteness of the metric $(26.3)_-$ with respect to time-like or null geodesics in the direction defined by $r = 2m$, $t' \to -\infty$ will now be irrelevant. We may remark that exact interior solutions corresponding to somehow realistic distributions of matter are not known yet. We shall derive in Section 29 one highly unrealistic interior solution corresponding to a material which is pressureless dust.

When the radius a of the star is larger than $2m$ the rate of contraction will depend on the equation of state. The contraction can eventually be stopped when $a > 2m$. Once, however, $a(t)$ has reached a value smaller than $2m$, there will be subsequently only a contraction, the rate of which cannot be smaller than a certain minimum determined by the form of the light cones (see Figure 9). It follows that, once the surface of the star has crossed the sphere $r = 2m$, the collapse will be completed in a finite interval of proper time. Remembering that according to Figure 9 no material particles or signals of any kind can cross the surface $r = 2m$ outwards, we see that an observer situated in the region $r > 2m$ cannot receive any information about what is happening in the central body when $a < 2m$. We express this fact by saying that the collapsed body constitutes a *black hole*.

After the collapse has been completed we shall have at $r = 0$ an intrinsic singularity of the metric due to an infinitely strong curvature. This can be seen in a qualitative way as follows. The average value of at least one component of the matter tensor $T_{\mu\nu}$ inside the star will be of the order of Mc^2/a^3. Because of the field Equation (19.7) the same will be true for the Ricci tensor $R_{\mu\nu}$ and consequently also for the Riemann tensor, of which the Ricci tensor is a contraction.

A possible risk in this reasoning is that a finite tensor may have some infinitely large components in an inappropriate frame. A proof can be obtained if one considers a scalar constructed from the Riemann tensor, e.g. the scalar

$$S = R^{\lambda\mu\nu\varrho} R_{\lambda\mu\nu\varrho} \,. \tag{27.1}$$

For the Schwarzschild field one finds that near $r=0$ the scalar S is of the order r^{-6}.

The gravitational collapse might in principle occur also in the Newtonian theory of gravitation: If the equation of state predicts an increase of the pressure which is slower than needed to halt the contraction, we shall get finally a point singularity at the center. Note that this singularity is of a much less drastic character than the one we have in the Schwarzschild solution. Indeed, in the newtonian case we only have the gravitational acceleration tending to an infinite value when $r \to 0$; the space is still euclidean and it is always possible to continue a straight line beyond the point $r=0$. On the contrary, in the Schwarzschild case the geometry degenerates at the point $r=0$ in such a way that it is impossible to continue a geodesic beyond this point.

Do black holes exist in nature? This is a fascinating question, but it will be extremely difficult to answer it by means of observations. It should be stressed that the theoretical prediction is of a qualitative character. It is based on the assumption of the spherical symmetry being maintained constantly during the contraction. This type of motion might however be unstable, leading finally to fragmentation rather than to collapse. The stability problem is very difficult to discuss, especially because for a reliable discussion we should know also the interior solution of the field equations. Moreover, it cannot be excluded a priori that entirely new phenomena may take place in extremely strong gravitational fields. We mention as an example the possibility of an emission of particles of the type of tachyons, which would invalidate the proof of the inevitability of gravitational collapse. Finally it has to be stressed that the prediction of gravitational collapse has been derived from the classical form of general relativity. The situation might be modified profoundly by quantum effects.

Needless to say that an observational proof of the existence of gravitational singularities and black holes would constitute an unprecedented upheaval of our system of physical concepts and ideas.

Exercises

V1: Complete the derivation of the form of the spherically symmetric metric using the Killing equation.

V2: Discuss in detail the identity (14.14) in the Schwarzschild case.

V3: In the Schwarzschild space-time consider a closed path C built by a null geodesic passing at a minimum distance $r_0 > 2m$ from the origin and a curve lying on the hypersurface $r = \text{const} \to \infty$. Let the vector k^μ be tangent to the null geodesic at some point P. Show that the light deflection is expressed by the variation δk^μ resulting from the parallel transport of k^μ along C.

V4: Derive the relation (24.9) for the light deflection by integrating Equation (9.6) over a 2-dimensional surface spanned by the closed path C of Exercise V3. (Assume $m/r_0 \ll 1$ and retain terms of the order of m/r_0 only.)

V5: Prove that the coordinate r is an affine parameter of the radial null geodesics of the metric (21.14) or (26.3).

V6: Prove that the time-like and the non-radial null directions of the metric $(26.3)_-$ have values $\mathrm{d}r/\mathrm{d}t'$ contained in the interval defined by (26.4):

$$-1 < \frac{\mathrm{d}r}{\mathrm{d}t'} < \frac{r-2m}{r+2m}.$$

V7: Study in some detail the mapping defined by Equations (26.7).

V8: Transform the Schwarzschild metric (21.14) into its *isotropic* form: Start with a transformation of the radial coordinate, $r = f(r')$; then transform to Cartesian-like coordinates:

$$x^1 = r' \sin\theta \cos\varphi, \qquad x^2 = r' \sin\theta \sin\varphi, \qquad x^3 = r' \cos\theta$$

and demand a metric of the form:

$$\mathrm{d}s^2 = \alpha(r')\,\mathrm{d}t^2 - \beta(r')\,\{(\mathrm{d}x^1)^2 + (\mathrm{d}x^2)^2 + (\mathrm{d}x^3)^2\}.$$

Determine $f(r')$ and then $\alpha(r')$ and $\beta(r')$.
(The final result is:

$$\mathrm{d}s^2 = \left(\frac{1 - m/2r'}{1 + m/2r'}\right)^2 \mathrm{d}t^2 - \left(1 + \frac{m}{2r'}\right)^4 \{(\mathrm{d}x^1)^2 + (\mathrm{d}x^2)^2 + (\mathrm{d}x^3)^2\}.)$$

CHAPTER VI

SOME OTHER EXACT SOLUTIONS

28. Fluid Without Pressure. Comoving Coordinates

The gravitational field inside the central body, which is the source of the Schwarzschild field, has to be determined as a solution of the inhomogeneous field equation (19.7). In the macroscopic description of matter the tensor $T^{\lambda\mu}$ is assumed to be of the form:

$$T^{\lambda\mu} = \varrho u^\lambda u^\mu + p^{\lambda\mu}\,; \tag{28.1}$$

ϱ is the density of proper mass, u^λ the macroscopic 4-dimensional velocity and $p^{\lambda\mu}$ the stress tensor. When dissipative effects can be neglected, we have the orthogonality relation:

$$p^{\lambda\mu} u_\mu = 0\,. \tag{28.2}$$

A simpler form of $T^{\lambda\mu}$ is the one for *perfect fluid.* In this case the stress tensor is isotropic, reducing to a scalar pressure p:

$$p^{\lambda\mu} = p\,(u^\lambda u^\mu - g^{\lambda\mu})\,. \tag{28.3}$$

The matter tensor is now of the form:

$$T^{\lambda\mu} = (\varrho + p)\, u^\lambda u^\mu - p g^{\lambda\mu}\,. \tag{28.4}$$

An oversimplified form of the matter tensor is the tensor $T^{\lambda\mu}$ which describes pressureless gas or simply *dust*:

$$T^{\lambda\mu} = \varrho u^\lambda u^\mu\,. \tag{28.5}$$

This form of $T^{\lambda\mu}$ is used sometimes in general relativity because of its simplicity, which makes the problem of solving the field equations easier. Note that a matter tensor of the form (28.5) can be a good approximation only in the case of a gas of extremely low density. Assuming the form (28.5) in the case of high densities is equivalent to the hypothesis that the material is a 'gas' with perfectly organised motions of its particles, u^μ being then the individual and not some average velocity of the particles. Evidently this case is highly unrealistic.

There are two known exact interior solutions for the spherically symmetric gravitational field. The first is a solution obtained by Schwarzschild. This is a time-independent solution referring to matter which is a perfect fluid of constant density, $\varrho = \text{const}$. In the case of a body like the Sun the assumption of constant density is unrealistic.

The second solution was given by Tolman. This is a time-dependent solution, the material being now a pressureless gas. In this case the material is falling freely in its own gravitational field and consequently cannot represent in any approximation the (very nearly) time-independent interior of the Sun. We shall discuss, however, here in detail the Tolman solution for the following two reasons:

(1) It will afford the occasion to develop and to use the important comoving coordinates.

(2) We shall obtain an intuitive, though oversimplified example of the gravitational collapse of an initially regular material distribution to a singularity.

The conservation law of energy and momentum is expressed in general relativity by Equation (19.5):

$$T^{\lambda\mu}{}_{;\mu} = 0. \tag{28.6}$$

This equation leads to some simple but important results when $T^{\lambda\mu}$ is of the form (28.5). Indeed, applying (28.6) to (28.5) we find:

$$(\varrho u^{\mu})_{;\mu} u^{\lambda} + \varrho u^{\mu} u^{\lambda}{}_{;\mu} = 0. \tag{28.7}$$

Multiplying this equation by u_{λ} and taking into account the relation (13.3):

$$u^{\lambda} u_{\lambda} = 1 \quad \rightarrow \quad u_{\lambda} u^{\lambda}{}_{;\mu} = 0,$$

we find:

$$(\varrho u^{\mu})_{;\mu} = 0. \tag{28.8}$$

Equation (28.7) reduces then to

$$u^{\lambda}{}_{;\mu} u^{\mu} = 0. \tag{28.9}$$

The meaning of this equation is, according to (13.12), that each particle of the dust moves on a geodesic. Equation (28.8) expresses the conservation of the rest-mass of the gas. Remembering (12.17) and (4.12) we see that we can write (28.8) also in the form:

$$(\varrho \sqrt{-g} u^{\mu})_{,\mu} = 0. \tag{28.10}$$

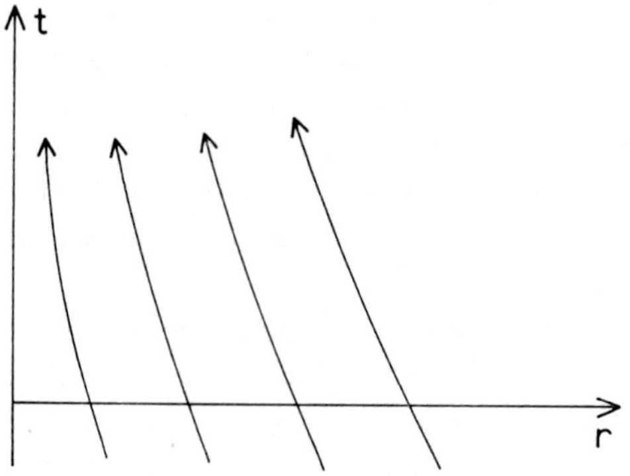

Fig. 12. Trajectories of particles in spherically symmetric motion.

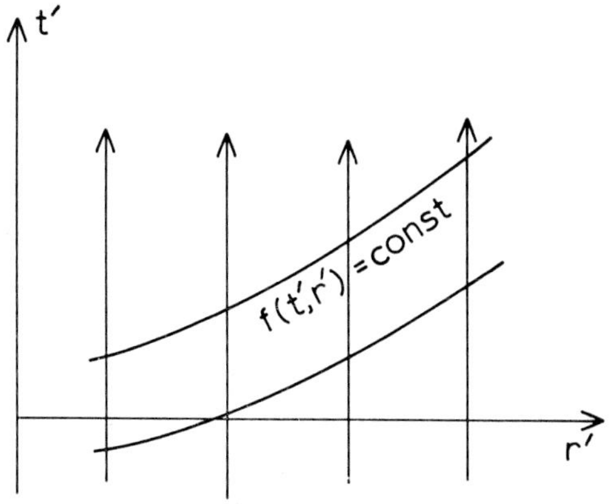

Fig. 13. Transforming to comoving coordinates.

A *comoving frame* has been found to be particularly well adapted to the present problem. The general idea is the following: Consider a system of particles moving so that their trajectories form a congruence of time-like non-intersecting curves filling a certain region of the space. Then we can introduce a frame x'^{μ} such that these curves are the parametric lines of the new coordinate $x'^0 = t'$.

We are interested here in the spherically symmetric case. In this case the particles have to move radially and therefore the motions can be described in the plane (t, r), as shown on Figure 12. The transformation to comoving coordinates is simply a transformation $(t, r) \to (t', r')$ – with the coordinates θ and φ unchanged – changing the trajectories of Figure 12 into lines parallel to the t'-axis as shown on Figure 13. According to a remark in Section 20 this transformation conserves the spherical symmetry. Hence we shall have a metric of the form (20.10):

$$ds^2 = -\alpha \, dr'^2 - \beta(d\theta^2 + \sin^2\theta \, d\varphi^2) + \gamma \, dt'^2 + \delta \, dt' \, dr', \qquad (28.11)$$

the coefficients α, β, γ and δ being functions of t' and r'.

The last term in (28.11) can be eliminated in the following way. We consider in the plane (t', r') curves $f(t', r') = \text{const}$ which are orthogonal, by the metric (28.11), to the curves $r' = \text{const}$. In the new coordinates

$$r'' = r', \qquad t'' = f(t', r')$$

we shall then have:

$$ds^2 = -\tilde{\alpha} \, dr''^2 - \beta(d\theta^2 + \sin^2\theta \, d\varphi^2) + \tilde{\gamma} \, dt''^2,$$

the lines $r'' = \text{const}$ being still the trajectories of the particles. This is the form of the metric which we shall use in the following discussion. We rewrite it using t, r instead of t'', r'' and α, γ instead of $\tilde{\alpha}$, $\tilde{\gamma}$:

$$ds^2 = \gamma \, dt^2 - \alpha \, dr^2 - \beta(d\theta^2 + \sin^2\theta \, d\varphi^2). \qquad (28.12)$$

We shall now use the result (28.9) according to which the trajectories of the particles are geodesics. Since on these geodesics r, θ and φ are constant, the velocity u^λ is of the form

$$u^\lambda = (u^0, 0, 0, 0). \tag{28.13}$$

The geodesic equation (13.11) reduces to:

$$\frac{du^\lambda}{ds} + \Gamma^\lambda_{00}(u^0)^2 = 0. \tag{28.14}$$

For $\alpha = i = 1, 2, 3$ this equation gives

$$\Gamma^i_{00} = 0 \quad \Leftrightarrow \quad g_{00,i} = 0.$$

Therefore

$$g_{00} = \gamma = \gamma(t). \tag{28.15}$$

We can now introduce a new time coordinate t' satisfying the condition

$$\sqrt{\gamma}\, dt = dt'$$

and in this way we obtain the final reduced form of the line element with $\gamma = 1$ (writing again t instead of t'):

$$ds^2 = dt^2 - e^\mu\, dr^2 - R^2(d\theta^2 + \sin^2\theta\, d\varphi^2); \tag{28.16}$$

$$\alpha = e^\mu, \qquad \beta = R^2,$$

μ and R being functions of t and r. With $g_{00} = 1$ we find for u^μ from (28.13), because of (13.3):

$$u^\mu = (1, 0, 0, 0) = u_\mu. \tag{28.17}$$

One verifies at once that the component $\lambda = 0$ of (28.14) is now satisfied identically. We recall that the lines r, θ, $\varphi = \text{const}$ are time-like geodesics and that t is the proper time on these geodesics.

The field equations are, from (19.7) and (28.5):

$$G_{\lambda\mu} \equiv R_{\lambda\mu} - \tfrac{1}{2} g_{\lambda\mu} R = -\kappa \varrho u_\lambda u_\mu.$$

According to (28.17) we have:

$$T_{00} = \varrho, \text{ all other components } T_{\lambda\mu} = 0. \tag{28.18}$$

The computation of the Einstein tensor $G_{\lambda\mu}$ is elementary but rather long. We give here the final result for the non-vanishing components G_{00}, G_{11}, $G_{22} = G_{33}/\sin^2\theta$ and G_{01}. The prime and the dot mean respectively derivatives with respect to r and t:

$$G_{00} = \frac{1}{R^2} e^{-\mu}(2RR'' + R'^2 - RR'\mu') - \frac{1}{R^2}(R\dot{R}\dot{\mu} + \dot{R}^2 + 1), \tag{28.19}$$

$$G_{11} = \frac{1}{R^2} e^\mu (2R\ddot{R} + \dot{R}^2 + 1) - \frac{1}{R^2} R'^2, \tag{28.20}$$

$$G_{22} = e^{-\mu}\left(- RR'' + \frac{R}{2} R'\mu' \right) + R\ddot{R} + \tfrac{1}{2}R^2\ddot{\mu} + \tfrac{1}{4}R^2\dot{\mu}^2 + \tfrac{1}{2}R\dot{R}\dot{\mu}, \quad (28.21)$$

$$G_{01} = 2\dot{R}' - R'\dot{\mu}. \quad (28.22)$$

The field equations corresponding to (28.20) and (28.22) are:

$$e^{\mu}(2R\ddot{R} + \dot{R}^2 + 1) - R'^2 = 0, \quad (28.23)$$

$$2\dot{R}' - R'\dot{\mu} = 0. \quad (28.24)$$

It can be verified without difficulty that the field equation corresponding to (28.21), $G_{22}=0$, is a consequence of (28.23) and (28.24) and can therefore be omitted. The field equation corresponding to (28.19) is, according to (28.18):

$$e^{-\mu}(2RR'' + R'^2 - RR'\mu') - (R\dot{R}\dot{\mu} + \dot{R}^2 + 1) = - \kappa\varrho R^2. \quad (28.25)$$

29. The Tolman Solution

We start with the simplest Equation (28.24). We have to demand $R'>0$, this following from the form (28.16) of the metric according to which the sphere $r=\text{const}$ has the area $4\pi R^2$. Consequently we can write (28.24) in the form:

$$\{2 \lg R' - \mu\}^{\cdot} = 0.$$

Therefore we have the integral

$$e^{\mu} = \frac{R'^2}{1+f}, \quad (29.1)$$

f being an arbitrary function of r.

The integration 'constant' $1+f(r)$ has been written in this form in order to simplify subsequent formulae. We must demand, in order to have the correct signature:

$$1+f>0, \quad \text{i.e.} \quad f>-1. \quad (29.2)$$

As one can see from (29.1) and (28.16), the vanishing of R' at some point means that the trajectories $r=\text{const}$ of two neighbouring points meet. This is equivalent to the formation of a caustic surface, on which the construction of the comoving coordinates would break down.

Equation (28.23) takes now the form:

$$2R\ddot{R} + \dot{R}^2 - f = 0. \quad (29.3)$$

Multiplying this equation by $\dot{R}$ we find:

$$(R\dot{R}^2 - fR)^{\cdot} = 0.$$

Therefore after integration:

$$\dot{R}^2 - f = \frac{F(r)}{R}, \tag{29.4}$$

$F(r)$ being the second integration constant.

Introducing in (28.25) the expression (29.1) for e^{μ} we find:

$$\frac{R}{R'}(f' - 2\dot{R}\dot{R}') - (\dot{R}^2 - f) = -\kappa\varrho R^2.$$

Using then Equation (29.4) we arrive at the relation:

$$\frac{F'}{R'} = \kappa\varrho R^2. \tag{29.5}$$

In order to complete the integration we have to integrate (29.4) once more and thus determine $R(t, r)$. The integration presents no difficulty, but there are 3 different cases which must be considered separately, as they lead to different analytic expressions. These cases correspond to $f>0$ or $f=0$ or $f<0$. We shall consider here only the case

$$f = 0 \tag{29.6}$$

in which the integration of (29.4) gives:

$$R = (\tfrac{3}{2})^{2/3} F^{1/3}\{T(r) - t\}^{2/3}, \tag{29.7}$$

$T(r)$ being a third integration constant.

We shall now consider Equation (29.5). From the form (28.16) of the metric we find, taking into account (29.1) and (29.6):

$$\sqrt{-g} = R'R^2 \sin\theta. \tag{29.8}$$

It follows that

$$\varrho R^2 R' = T_{00}\sqrt{-g}/\sin\theta.$$

From (28.10) and (28.17) we find now:

$$(\varrho R^2 R')^{\cdot} = 0. \tag{29.9}$$

This relation is in agreement with (29.5) because $F=F(r)$. Actually we can consider $F(r)$ as determining the distribution of proper mass or of energy inside the central body by means of Equation (29.5) written in the form:

$$F' = \kappa\varrho\sqrt{-g}/\sin\theta = \kappa T_{00}\sqrt{-g}/\sin\theta. \tag{29.10}$$

The meaning of the function $f(r)$ can be found from Equation (29.4), if we notice that $R(t, r)$ can be considered as defining a certain *distance* of the point (t, r) from the centre of symmetry. This is based on the remark that the sphere $r, t=\text{const}$ has the area $4\pi R^2$. But then $\dot{R}$ is the (radial) velocity of the particle at (t, r). Equation (29.4) expresses now the fact that, when $f=0$, the particle has the velocity $\dot{R}=0$ at $R=\infty$.

This is therefore the case which in the Newtonian theory is called the *parabolic* motion of the particles. In this case we have integrated the field equations completely. If $f>0$, the particles can again reach $R=\infty$, having there a velocity $\dot{R}\neq 0$: This is the case of the *hyperbolic* motion. On the contrary, if $f<0$, the particle can only reach a certain maximum distance $R_{\max}$ determined by the relation:

$$F/R_{\max} = -f,$$

at which the sign of the velocity must change: maximum expansion turning into contraction. This is the case of the *elliptic* motion of the particles.

The function $T(r)$ has no direct physical meaning. It can also be changed by a transformation of the form $r\to r'(r)$, which conserves the form (28.16) of the metric. For example, if $T(r)$ is an increasing function of r we can bring it into the 'standard' form $T(r)=r$.

We mentioned already that we have to demand $R'>0$: The area of the 2-dimensional sphere t, $r=\text{const}$ must increase with r. We then see from (29.5) that we must demand $F'\geqslant 0$ for a non-negative density ϱ. Thus F will increase with r in the interior of the dust distribution and remain constant outside. From Equation (29.7) we derive the relation:

$$\frac{3R'}{R} = \frac{F'}{F} + \frac{2T'}{T-t}. \tag{29.11}$$

From this relation we conclude that in order to have $R'>0$ also outside the dust distribution, where we have $F'=0$, we must demand $T'\neq 0$. If we choose as an example

$$T(r) = r, \tag{29.12}$$

then we shall have

$$\frac{T'}{T-t} = \frac{1}{r-t} > 0 \quad \text{for} \quad r > t. \tag{29.13}$$

This solution will therefore be acceptable only in the half-plane (t, r) situated below the line L (Figure 14) defined by the equation

$$t = T(r) = r.$$

We shall now assume that $F'=0$, i.e. $F=\text{const}\equiv F_0$ for $r>a$ (Figure 14). It follows then from the Birkhoff theorem that the metric (28.16) is, for $r>a$, just another form of the Schwarzschild metric (21.14). We shall verify this conclusion by determining the transformation $(t, r)\to(t', r')$ which will bring the metric (28.16) with $F=F_0$ into some already known form of the Schwarzschild metric. We shall consider only the case in which $T(r)$ is given by (29.12).

For $r>a$ we find from (29.7) and (29.12), with $F=F_0$:

$$R^3 = \tfrac{9}{4}F_0(r-t)^2. \tag{29.14}$$

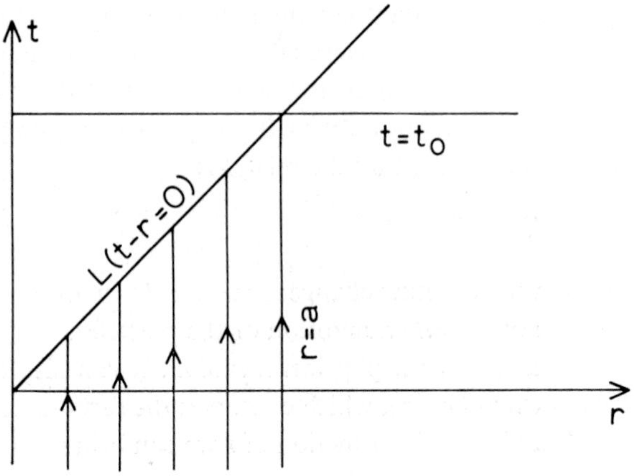

Fig. 14. Gravitational collapse of a cloud of dust.

Therefore:

$$R' = \left(\frac{2}{3}\frac{F_0}{r-t}\right)^{1/3} \tag{29.15}$$

and the metric (28.16) takes, because of (29.1), the form:

$$ds^2 = dt^2 - \left\{\frac{2F_0}{3(r-t)}\right\}^{2/3} dr^2 - \{\tfrac{9}{4}F_0 (r-t)^2\}^{2/3} (d\theta^2 + \sin^2\theta \, d\varphi^2). \tag{29.16}$$

We consider now the transformation $(t, r) \to (t', r')$ determined by the following equations:

$$t = t' + \int \frac{\alpha \, dr'}{1+\alpha}, \qquad r = t' + \int \left(\frac{1}{\alpha} + \frac{\alpha}{1+\alpha}\right) dr'; \qquad \alpha(r') = \sqrt{\frac{F_0}{r'}}. \tag{29.17}$$

The difference $r-t$ is given by a simple expression:

$$r - t = \int \frac{1}{\alpha} dr' = \frac{2}{3}\left(\frac{r'^3}{F_0}\right)^{1/2}. \tag{29.18}$$

The expressions for t and r are more complicated and we shall not determine them, as we do not need them explicitly. From (29.17) we find:

$$dt = dt' + \frac{\alpha}{1+\alpha} dr', \qquad dr = dt' + \left(\frac{1}{\alpha} + \frac{\alpha}{1+\alpha}\right) dr'. \tag{29.19}$$

Introducing these expressions in (29.16) we find after some elementary calculations:

$$\left.\begin{aligned} ds^2 = \left(1 - \frac{F_0}{r'}\right) dt'^2 - \left(1 + \frac{F_0}{r'}\right) dr'^2 - \\ - \frac{2F_0}{r'} dt' \, dr' - r'^2 (d\theta^2 + \sin^2\theta \, d\varphi^2). \end{aligned}\right\} \tag{29.20}$$

This is the form $(26.3)_-$ of the Schwarzschild metric, provided we put

$$F_0 = 2m\,. \tag{29.21}$$

Equation (29.18) shows that the line L of Figure 14, defined by the equation $r-t=0$, is the worldline of the centre of symmetry. With this remark we see that the meaning of Figure 14 is the following. The particles of the cloud of dust, which constitute the source of the gravitational field, move on the geodesics $r=\text{const}<a$. Each of these particles reaches the centre of symmetry at the moment $t=r$. At $t=t_0$ all particles have reached the centre of symmetry and the gravitational collapse has been completed. For $t>t_0$ we have everywhere empty space except at the points of the line L, where we have the singularity. Thus the diagram of the Figure 14 describes, together with the metric (28.16), the formation of a black hole by the gravitational collapse of a cloud of dust.

We shall mention only briefly the case in which the function $T(r)$ is decreasing monotonically, so that it can be put into the form $T(r)=-r$. If we have again $F=\text{const}=F_0$ for $r>a$, it follows from (29.11) that in order to have $R'>0$ also for $r>a$ we must consider only the points of the plane (t, r) which are above the line L' determined by the equation

$$t = T(r) = -r\,.$$

In this case all particles of the dust move outwards, the particle corresponding to the value r starting from the centre at the moment $t=-r$. A transformation similar to (29.17) brings this metric into the form $(26.3)_+$. This is, therefore, the description in comoving coordinates of a 'white hole', a cloud of expanding dust coming out of the singularity.

We end this discussion with the following two remarks. In the comoving frame it has been possible to determine at the same time the gravitational field inside as well as outside its material sources. Indeed, the metric (28.16) determines with $F'\neq 0$ the interior and with $F'=0$ the exterior field. Note, however, that the metric is now time-dependent everywhere, the comoving coordinates t and r being less intuitive than those used in (21.14) or (26.3).

The last remark concerns a distribution of dust, the particles of which have vanishing rest-mass (being e.g. photons). The trajectories of the particles will then be null-geodesics and it is easy to see that the metric cannot be diagonal in comoving coordinates. The solution of this problem is extremely simple. The metric has exactly the form (26.3), the only difference being that now the quantity m is not a constant, but an arbitrary function of $t'\mp r$. This is the *Vaidya solution* which will be proposed as an exercice at the end of this chapter.

30. The Kerr Solution

This is an important generalization of the Schwarzschild solution. It describes the gravitational field of a material source at rest having mass *and* angular momentum.

The angular momentum determines a physically significant direction. Consequently the field cannot be spherically symmetric: It has only rotational or *axial* symmetry, the axis of symmetry having the direction of the angular momentum. The field is also time-independent and the metric will therefore have in this case two Killing vectors, one timelike and one space-like, corresponding respectively to the time-independence and to the axial symmetry of the field. We shall use a coordinate system adapted to both these Killing vectors; the space coordinates will be the polar coordinates r, θ and φ, the axis $\theta=0$ coinciding with the axis of symmetry. Therefore the metric will be independent of the time-coordinate $x^0 \equiv t$ and of the angle $x^3 \equiv \varphi$ around the axis of symmetry.

Kerr obtained his metric in an elegant manner using results and theorems about algebraically special metrics (see end of Section 15). As these results are not treated in these lectures, we shall give the Kerr metric without derivation. It is the following:

$$\begin{aligned} \mathrm{d}s^2 = {} & (1-X)\,\mathrm{d}t^2 - (1+X)\,\mathrm{d}r^2 - 2X\,\mathrm{d}t\,\mathrm{d}r - (r^2 + a^2\cos^2\theta)\,\mathrm{d}\theta^2 \\ & - (r^2 + a^2 + a^2\sin^2\theta\, X)\sin^2\theta\,\mathrm{d}\varphi^2 - 2a\sin^2\theta\, X\,\mathrm{d}t\,\mathrm{d}\varphi \\ & - 2a\sin^2\theta(1+X)\,\mathrm{d}r\,\mathrm{d}\varphi\,; \end{aligned} \tag{30.1}$$

$$X = \frac{2mr}{r^2 + a^2\cos^2\theta}. \tag{30.2}$$

The metric depends on the two constants m and a. One verifies at once that with $a=0$ the metric (30.1) reduces to the Schwarzschild metric in the form $(26.3)_-$. The metric (30.1) is a solution of the vacuum field equation (19.4). We only mention here, without a proof, that the metric (30.1) is of type D (see Section 15).

The diagonal components of the metric (30.1) are seen at once to tend to the Minkowski values (in polar coordinates) when $r \to \infty$. The non-diagonal terms are of an unusual form. It is therefore necessary, in order to prove that the metric (30.1) tends to the Minkowski metric when $r \to \infty$, to transform it to Cartesian-type coordinates. This is obtained by the transformation:

$$(r - ia)\,e^{i\varphi}\sin\theta = x + iy, \qquad r\cos\theta = z. \tag{30.3}$$

The resulting new form of the metric is:

$$\left.\begin{aligned} \mathrm{d}s^2 = {} & \mathrm{d}t^2 - (\mathrm{d}x^2 + \mathrm{d}y^2 + \mathrm{d}z^2) - \frac{2mr}{(r^4 + a^2z^2)(r^2+a^2)^2} \times \\ & \times \{r^2(x\,\mathrm{d}x + y\,\mathrm{d}y) + ar(x\,\mathrm{d}y - y\,\mathrm{d}x) + \\ & + (r^2 + a^2)(z\,\mathrm{d}z + r\,\mathrm{d}t)\}^2. \end{aligned}\right\} \tag{30.4}$$

From (30.3) we find:

$$R^2 \equiv x^2 + y^2 + z^2 = r^2 + a^2\sin^2\theta. \tag{30.5}$$

Therefore, when $a/r \ll 1$:

$$R = r + \frac{a^2\sin^2\theta}{2r} + \cdots.$$

It follows then from (30.4) that $g_{\lambda\mu} \to \eta_{\lambda\mu}$ when $R \to \infty$, $\eta_{\lambda\mu}$ being the diagonal Minkowski metric in Cartesian coordinates:

$$\eta_{\lambda\mu} = (1, -1, -1, -1).$$

We also remark that the metric (30.1) is singular at $r=0=\cos\theta$. This is expressed in the new coordinates x, y, z by the relations:

$$x^2 + y^2 = a^2, \qquad z = 0;$$

i.e. the singularity of the metric (30.4) is the ring of radius a on the plane $z=0$.

We can determine from (30.4) the next order terms in the development in $1/R$ of the components $g_{0\lambda}$. The interesting terms are the following:

$$\mathrm{d}s^2 = \left(1 - \frac{2m}{R} + \cdots\right) \mathrm{d}t^2 - \frac{4ma}{R^3}(x\,\mathrm{d}y - y\,\mathrm{d}x)\,\mathrm{d}t + \cdots. \tag{30.6}$$

We shall see later, from the discussion of the conservation laws in Section 35 and also from the discussion of the weak gravitational fields in Section 32, that the $1/R$-term in g_{00} determines the total mass of the field, while the $1/R^2$-terms in g_{0i} ($i=1, 2, 3$) determine the components of the angular momentum. The final result is the following: The metric (30.1) or (30.4) describes a field of which the total mass is m and the angular momentum (in the direction of the axis of symmetry) am.

What kind of a material body could be the source of the Kerr field? To answer this question one should determine the continuation of the Kerr metric into the interior of an appropriate body, but this has not been possible until now. At present one usually assumes that the Kerr metric is valid everywhere outside of the singularities of (30.1). In this form it plays an important role in the general discussion of the gravitational collapse and of black holes. We shall discuss it here briefly in order to show that it has an horizon and consequently it is a black hole.

In the Schwarzschild metric the horizon is determined by the equation:

$$g_{00} = 0. \tag{30.7}$$

In the Kerr metric this equation has, from (30.1), the form:

$$r^2 + a^2\cos^2\theta - 2mr = (r-m)^2 - (m^2 - a^2\cos^2\theta) = 0. \tag{30.8}$$

We shall assume

$$a < m, \tag{30.9}$$

in which case Equation (30.8) is equivalent to:

$$r = m \pm \sqrt{m^2 - a^2\cos^2\theta}. \tag{30.10}$$

Therefore, Equation (30.8) determines two surfaces S_+ and S_-, having in a meridian plane and in the coordinates x, y, z the form shown on Figure 15. However, when $a \neq 0$, these two surfaces are not characteristic; the normal to the surfaces (30.10)

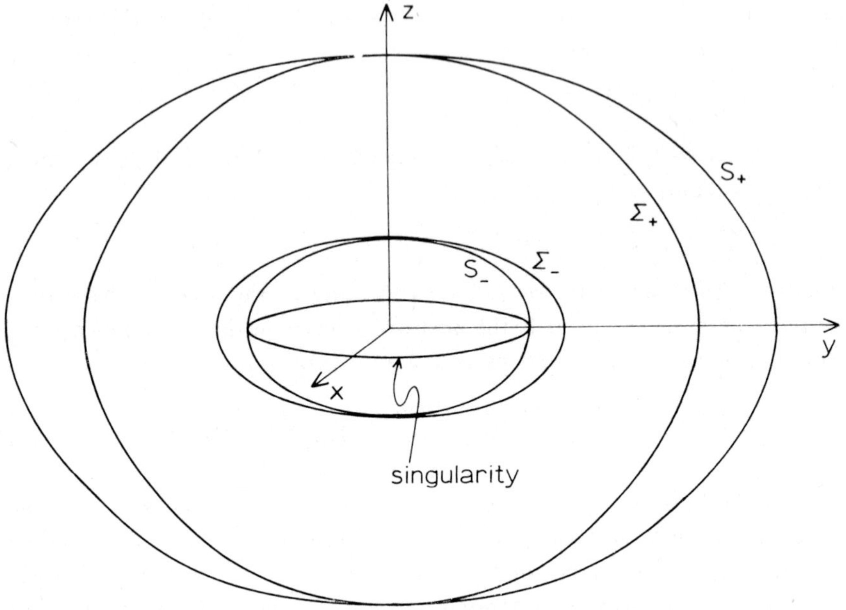

Fig. 15. The horizon Σ_+ and the surfaces Σ_-, S_+ and S_- of the Kerr metric on the meridian plane (y, z). The ring singularity is also shown.

is not a null-vector. This is verified at once with the help of the following values of $g^{\lambda\mu}$ determined from (30.1):

$$g^{00} = 1 + X, \qquad g^{11} = X - \frac{r^2 + a^2}{r^2 + a^2\cos^2\theta}, \qquad g^{01} = -X;$$

$$g^{22} = \frac{-1}{r^2 + a^2\cos^2\theta}, \qquad g^{33} = \frac{-1}{(r^2 + a^2\cos^2\theta)\sin^2\theta}, \qquad g^{13} = \frac{a}{r^2 + a^2\cos^2\theta}. \tag{30.11}$$

The meaning of the surfaces S_+ and S_- is the following. The Killing vector corresponding to the time-independence of the metric (30.1) has the direction of the t-axis. This vector is time-like only outside S_+ or inside S_-; it is a null vector on S_+ or S_- and space-like between S_+ and S_-. It follows then from a qualitative reasoning that we should expect an horizon somewhere between S_+ and S_-.

This is really the case as one can prove by discussing the equation

$$r^2 + a^2 - 2mr = (r - m)^2 - (m^2 - a^2) = 0. \tag{30.12}$$

This is equivalent to

$$r = m \pm \sqrt{m^2 - a^2}, \tag{30.13}$$

determining two surfaces Σ_+ and Σ_- of the form shown on Figure 15 (in the coordinates

x, y, z). The normal to these surfaces is the vector

$$p_\lambda = (0, 1, 0, 0), \tag{30.14}$$

which is a null vector, as one verifies at once using the expression for g^{11} given in (30.11). Thus the surfaces Σ_+ and Σ_- are characteristic.

The surface Σ_+ is the horizon of the Kerr metric. This follows from the properties of characteristic surfaces (see Section 47) combined with the remark that in the Kerr metric the vector

$$k^\mu = (1, -1, 0, 0) \tag{30.15}$$

is a null vector. Since the vector k^μ is future-pointing ($k^0 > 0$) and directed towards decreasing values of r ($k^1 < 0$), the future light-cone of any point of Σ_+ lies inside Σ_+. Consequently there are no outgoing null or time-like directions and Σ_+ is the horizon.

The discussion of the geodesics or equivalently of the motion of test particles in the Kerr metric is essentially more complicated than in the Schwarzschild metric. We only mention here that it is not necessary now for any particle crossing the horizon to fall on the singularity. Qualitatively this is a consequence of the existence of the interior of S_- (Figure 15), in which the direction of the t-axis is again time-like. As the Schwarzschild metric, the Kerr metric also is incomplete. The extension to a complete manifold by the use of Kruskal-type coordinates has been discussed and leads to results which are still more puzzling, from the physical point of view, than those found in the Schwarzschild case.

As a last remark we mention that applying a time inversion to the metric (30.1) we obtain a new metric which is related to (30.1) in the same way as the Schwarzschild metric $(26.3)_+$ is related to $(26.3)_-$.

Exercises

VI1: Discuss the Tolman solution in the special case corresponding to $T(r) = \text{const}$.

VI2: Determine the horizon of the Schwarzschild solution expressed in comoving coordinates.

VI3: *The Vaidya solution*: Consider a metric of the form $(26.3)_+$ assuming that m is not a constant, but an arbitrarily given function of the retarded time $u = t' - r$, and compute the corresponding Ricci tensor $R_{\mu\nu}$. Then determine the matter tensor $T_{\mu\nu}$ from the field equations (19.7) and prove that it has the form:

$$T_{\mu\nu} = \varrho k_\mu k_\nu,$$

k_μ being a null vector.

VI4: Prove that the equation

$$r = m + \sqrt{m^2 - a^2} \cos(\theta + \lambda), \qquad \lambda = \text{const.},$$

determines a characteristic hypersurface Σ of the Kerr metric. Show that Σ is tangent to Σ_+ and to Σ_-, having a common bicharacteristic with each one of these surfaces. (Use the result of exercice XI3.)

CHAPTER VII

WEAK GRAVITATIONAL FIELDS

31. The Linear Approximation

As we mentioned already in the beginning of Section 26, the gravitational field outside the Sun is described by a metric which differs very little from the Minkowski metric: The difference is expressed by the term $2m/r$ of the Schwarzschild metric, having its maximum value of the order of 10^{-5} at the surface of the Sun. Inside the Sun this difference will be larger, depending on the detailed distribution of the Sun's mass, but it is not expected to exceed the order 10^{-4} (see Section 32).

A gravitational field described by a metric which differs very little from the Minkowski metric is called a *weak* gravitational field. According to the preceding remarks the gravitational field of the Sun is weak everywhere. Actually this is the case for nearly all stellar objects. White dwarfs may have a radius up to 100 times smaller than the radius of the Sun; the difference from the Minkowski metric will still be of the order 10^{-2} at the centre of the star. The only exception, in which the gravitational field will not be weak, is the case of a neutron star (pulsar), the radius of which is of the order of $2m$.

It follows from these remarks that weak gravitational fields deserve special attention. We shall develop in the following an approximation method applicable to them.

The problem is simplified if we use Cartesian-like coordinates in the 3-dimensional space. The Minkowski metric is then diagonal, having the simple form (15.4):

$$\eta_{\mu\nu} = \eta^{\mu\nu} = (1, -1, -1, -1). \tag{31.1}$$

The definition of a weak gravitational field is now straightforward: The gravitational field described by the metric $g_{\mu\nu}$ is weak if the difference $g_{\mu\nu} - \eta_{\mu\nu}$ is small compared with one, i.e. if:

$$|g_{\mu\nu} - \eta_{\mu\nu}| \ll 1. \tag{31.2}$$

We shall assume that the condition (31.2) is satisfied in the entire space-time. Note that the condition (31.2) can only be satisfied in a certain class of coordinate systems which correspond approximately to the inertial frames of the Minkowski space.

We may remark that a gravitational field which is not weak everywhere, as e.g. the field of a neutron star, can be strong only in a certain tube-like region of the space corresponding to the neighbourhood of the field-producing body. Outside this region the field will be weak, as it will satisfy the condition (31.2) because of the general demand of an asymptotically flat space:

$$g_{\mu\nu} \to \eta_{\mu\nu} \quad \text{for} \quad r \to \infty. \tag{31.2'}$$

The general approximation method for weak fields has to be based on an appropriate small parameter which we shall identify later. We shall assume that $g_{\mu\nu}$ is developped in a series of the form:

$$g_{\mu\nu} = \eta_{\mu\nu} + g'_{\mu\nu} + g''_{\mu\nu} + \cdots ; \tag{31.3}$$

$g'_{\mu\nu}$ is the term of first order, $g''_{\mu\nu}$ of second and so on. Here we shall limit ourselves to the first approximation, writing instead of (31.3):

$$g_{\mu\nu} = \eta_{\mu\nu} + h_{\mu\nu}, \qquad h_{\mu\nu} \equiv g'_{\mu\nu}. \tag{31.4}$$

We omit all subsequent terms in (31.3) and we shall retain in the field equation only the terms which are linear in $h_{\mu\nu}$ or derivatives of $h_{\mu\nu}$. It is for this reason that this approximation is called the *linear* approximation and the corresponding field equation, which we shall derive now, the *linearized* field equation. We may mention here that higher approximations have been considered only in some special cases. They are much more complicated and it is not possible to discuss them generally, as we shall be able to do for the linear approximation.

The linearized field equation is obtained without difficulty. We first calculate the Christoffel symbols from the general formula (12.9). The contravariant tensor $g^{\mu\nu}$ is of the form:

$$g^{\mu\nu} = \eta^{\mu\nu} - h^{\mu\nu}. \tag{31.5}$$

The quantity $h^{\mu\nu}$ will be determined from the relation (11.8):

$$g^{\lambda\nu} g_{\mu\nu} = (\eta^{\lambda\nu} - h^{\lambda\nu})(\eta_{\mu\nu} + h_{\mu\nu}) = \delta^{\lambda}_{\mu}.$$

We find in the linear approximation:

$$\eta^{\lambda\nu} h_{\mu\nu} - \eta_{\mu\nu} h^{\lambda\nu} = 0.$$

Multiplying this relation by $\eta^{\varrho\mu}$ we obtain:

$$h^{\lambda\varrho} = \eta^{\lambda\nu} \eta^{\varrho\mu} h_{\nu\mu}. \tag{31.6}$$

Introducing (31.4) and (31.5) in (12.9) we find in the linear approximation:

$$\Gamma^{\lambda}_{\mu\nu} = \tfrac{1}{2} \eta^{\lambda\varrho} (h_{\varrho\nu,\mu} + h_{\mu\varrho,\nu} - h_{\mu\nu,\varrho}). \tag{31.7}$$

We then calculate the Riemann tensor from (14.2). The omitted terms are of the second order and so we have in the linear approximation:

$$R_{\lambda\mu\nu\varrho} = \tfrac{1}{2} (h_{\mu\nu,\lambda\varrho} + h_{\lambda\varrho,\mu\nu} - h_{\lambda\nu,\mu\varrho} - h_{\mu\varrho,\lambda\nu}). \tag{31.8}$$

Finally we calculate the Ricci tensor by contracting the Riemann tensor (31.8). In the linear approximation we have to multiply (31.8) by $\eta^{\lambda\varrho}$. The final result is:

$$R_{\mu\nu} = \tfrac{1}{2} \Box h_{\mu\nu} - \tfrac{1}{2} \eta^{\lambda\varrho} (\gamma_{\lambda\mu,\nu\varrho} + \gamma_{\lambda\nu,\mu\varrho}). \tag{31.9}$$

The symbol $\square$ is the D'Alembert operator in the Minkowski space:

$$\square\, h_{\mu\nu} = \eta^{\lambda\varrho} h_{\mu\nu,\,\lambda\varrho}, \tag{31.10}$$

and the quantities $\gamma_{\lambda\mu}$ are the following linear combinations of $h_{\lambda\mu}$:

$$\gamma_{\lambda\mu} = h_{\lambda\mu} - \tfrac{1}{2}\eta_{\lambda\mu} h, \qquad h \equiv \eta^{\varrho\sigma} h_{\varrho\sigma}. \tag{31.11}$$

In order to derive the linearized field equation we have to calculate also the scalar R to first order. This is found at once if we multiply (31.9) by $\eta^{\mu\nu}$:

$$R = \eta^{\mu\nu} R_{\mu\nu} = \tfrac{1}{2}\square\, h - \eta^{\lambda\varrho}\eta^{\mu\nu}\gamma_{\lambda\mu,\,\nu\varrho}. \tag{31.12}$$

The linearized field equation follows from (19.7):

$$R_{\mu\nu} - \tfrac{1}{2}\eta_{\mu\nu} R = -\,\kappa T_{\mu\nu}.$$

Introducing in this equation the expressions (31.9) and (31.12) we find finally:

$$\square\, \gamma_{\mu\nu} - \eta^{\lambda\varrho}(\gamma_{\lambda\mu,\,\nu\varrho} + \gamma_{\lambda\nu,\,\mu\varrho}) + \eta_{\mu\nu}\eta^{\lambda\varrho}\eta^{\alpha\beta}\gamma_{\lambda\alpha,\,\varrho\beta} = -\,2\kappa T_{\mu\nu}. \tag{31.13}$$

It can be verified directly that, if we apply on the left-hand side of (31.13) the operator $\eta^{\lambda\nu}(\partial/\partial x^\lambda)$, the result is identically zero. Therefore $T_{\mu\nu}$ must satisfy the condition:

$$\eta^{\lambda\nu} T_{\mu\nu,\,\lambda} = 0. \tag{31.14}$$

Equation (31.14) is the conservation law of energy and momentum in special relativity. This remark clarifies the physical meaning of the field equation (31.13): We have to start with the matter tensor $T_{\mu\nu}$ which describes the physical system under consideration in special relativity, i.e. disregarding the gravitational field. This matter tensor is then the source of the gravitational field in the linear approximation. The gravitational field is weak because of the smallness of the gravitational constant κ (in laboratory units). Therefore κ is the small parameter used implicitely in the development (31.3). In order to indicate this situation we shall write in the following ${}_0T_{\mu\nu}$ instead of $T_{\mu\nu}$: The matter tensor is of order zero; the right-hand side of (31.13) is of the first order – as is also the left-hand side of this equation – because of the factor κ.

Let us now consider a coordinate transformation of the form:

$$\tilde{x}^\mu = x^\mu + \varepsilon\xi^\mu(x^\nu), \tag{31.15}$$

ε being of the order of $h_{\mu\nu}$. We shall then have:

$$g_{\mu\nu} = \eta_{\mu\nu} + h_{\mu\nu} = \frac{\partial\tilde{x}^\varrho}{\partial x^\mu}\frac{\partial\tilde{x}^\sigma}{\partial x^\nu}\tilde{g}_{\varrho\sigma} = (\delta^\varrho_\mu + \varepsilon\xi^\varrho{}_{,\,\mu})(\delta^\sigma_\nu + \varepsilon\xi^\sigma{}_{,\,\nu})\,\tilde{g}_{\varrho\sigma}.$$

Putting

$$\tilde{g}_{\varrho\sigma} = \eta_{\varrho\sigma} + \tilde{h}_{\varrho\sigma} \tag{31.16}$$

and retaining only terms of the first order we find:

$$h_{\mu\nu} = \tilde{h}_{\mu\nu} + \varepsilon(\eta_{\mu\varrho}\xi^\varrho{}_{,\,\nu} + \eta_{\varrho\nu}\xi^\varrho{}_{,\,\mu}). \tag{31.17}$$

This relation shows that $\tilde{h}_{\mu\nu}$ and $h_{\mu\nu}$ are of the same order of magnitude. Consequently, we can use the linearized field equation (31.13) in the coordinate system x^μ as well as in $\tilde{x}^\mu$.

From (31.17) we derive a relation between $\gamma_{\mu\nu}$ and the corresponding quantity $\tilde{\gamma}_{\mu\nu}$,

$$\tilde{\gamma}_{\mu\nu} = \tilde{h}_{\mu\nu} - \tfrac{1}{2}\eta_{\mu\nu}\tilde{h}\,, \qquad \tilde{h} = \eta^{\varrho\sigma}\tilde{h}_{\varrho\sigma}\,.$$

This relation is:

$$\gamma_{\mu\nu} = \tilde{\gamma}_{\mu\nu} + \varepsilon\left(\eta_{\mu\varrho}\xi^{\varrho}{}_{,\,\nu} + \eta_{\nu\varrho}\xi^{\varrho}{}_{,\,\mu} - \eta_{\mu\nu}\xi^{\varrho}{}_{,\,\varrho}\right). \tag{31.18}$$

Multiplying this relation by $\eta^{\lambda\nu}(\partial/\partial x^\lambda)$ we find:

$$\eta^{\lambda\nu}\gamma_{\mu\nu,\,\lambda} = \eta^{\lambda\nu}\tilde{\gamma}_{\mu\nu,\,\lambda} + \varepsilon\eta_{\mu\varrho}\,\Box\,\xi^{\varrho}\,. \tag{31.19}$$

Equation (31.19) allows an essential simplification of the field equation (31.13). Indeed, supposing that $\gamma_{\mu\nu}$ is given, we can apply a transformation of the type (31.15) with a ξ^μ satisfying the equation:

$$\eta^{\lambda\nu}\gamma_{\mu\nu,\,\lambda} = \varepsilon\eta_{\mu\varrho}\,\Box\,\xi^{\varrho}\,. \tag{31.20}$$

Since this equation is inhomogeneous wave equation in Minkowski space, we can always determine solutions $\varepsilon\xi^\varrho$ for any given $\gamma_{\mu\nu}$. But then we shall have from (31.19):

$$\eta^{\lambda\nu}\tilde{\gamma}_{\mu\nu,\,\lambda} = 0\,. \tag{31.21}$$

In the coordinate system $\tilde{x}^\mu$ the linearized field equation is Equation (31.13) with $\gamma_{\mu\nu}$ replaced by $\tilde{\gamma}_{\mu\nu}$. Because of (31.21) we now have a much simpler equation:

$$\Box\,\tilde{\gamma}_{\mu\nu} = -\,2\kappa\,{}_0T_{\mu\nu}\,. \tag{31.22}$$

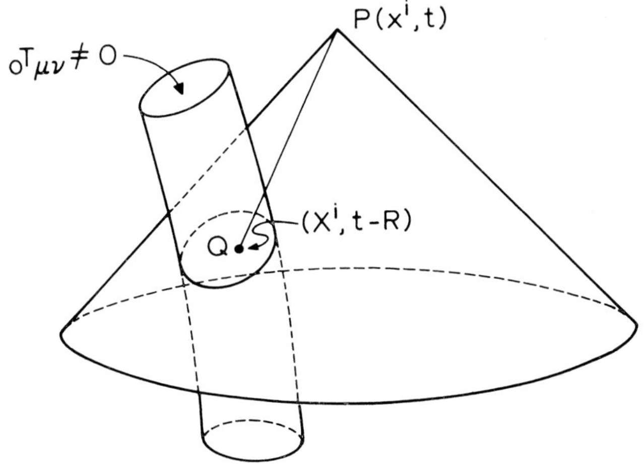

Fig. 16. Determining the retarded field at a point P.

Note the analogy of the field equation (31.22), valid under the *coordinate condition* (31.21), to Maxwell's equation with the Lorentz gauge condition:

$$\Box A_\mu = s_\mu, \qquad \eta^{\lambda\mu} A_{\mu,\lambda} = 0. \tag{31.23}$$

Thus the coordinate condition (31.21) is for the gravitational field what the Lorentz gauge is for the electromagnetic field. We shall write in the following $\gamma_{\mu\nu}$ instead of $\tilde{\gamma}_{\mu\nu}$.

The analogy between (31.22) and the Maxwell equation (31.23) allows to write down immediately the expression giving $\gamma_{\mu\nu}$ when the sources ${}_0T_{\mu\nu}$ are known. The $\gamma_{\mu\nu}$ will be the *retarded solution* of (31.22):

$$\gamma_{\mu\nu}(x^i, t) = -\frac{2\kappa}{4\pi} \int \frac{1}{R}\, {}_0T_{\mu\nu}(X^i, t - R)\, \mathrm{d}^3X; \tag{31.24}$$

$$R^2 = \sum_i (x^i - X^i)^2, \qquad i = 1, 2, 3. \tag{31.25}$$

We recall that the integration in (31.24) is made over the (minkowskian) past light cone of the point P having the coordinates x^i and t (Figure 16).

32. Applications. Mass and Angular Momentum

A1. Total mass in the static case

We consider first a distribution of matter at equilibrium described in special relativity by the following simple ${}_0T_{\mu\nu}$:

$${}_0T_{00} = \varrho c^2, \qquad {}_0T_{\mu\nu} = 0 \quad \text{for} \quad \mu\nu \neq 00, \tag{32.1}$$

the density ϱ being independent of the time-coordinate x^0. One verifies at once that the conservation law (31.14) is satisfied by the tensor (32.1).

We assume that this is the only source of the gravitational field which must, therefore, be static. Consequently the field equation (31.22) reduces to:

$$\Delta\gamma_{\mu\nu} = 2\kappa\, {}_0T_{\mu\nu}, \tag{32.2}$$

where Δ is the Laplace operator. With the matter tensor (32.1) we find:

$$\Delta\gamma_{00} = 2\kappa\varrho c^2, \qquad \Delta\gamma_{\mu\nu} = 0 \quad \text{for} \quad \mu\nu \neq 00. \tag{32.3}$$

If we write the field equation of the Newtonian theory of gravitation,

$$\Delta\varphi = -4\pi G\varrho, \tag{32.4}$$

we see that the solution of (32.3) is:

$$\gamma_{00} = \frac{-\kappa c^2}{2\pi G}\varphi, \qquad \gamma_{\mu\nu} = 0 \quad \text{for} \quad \mu\nu \neq 00. \tag{32.5}$$

In order to determine $h_{\mu\nu}$ we multiply (31.11) by $\eta^{\mu\nu}$:

$$\eta^{\mu\nu}\gamma_{\mu\nu} \equiv \gamma = h - \tfrac{4}{2}h = -h.$$

We can therefore solve (31.11) with respect to $h_{\mu\nu}$. The result is:

$$h_{\mu\nu} = \gamma_{\mu\nu} - \tfrac{1}{2}\eta_{\mu\nu}\gamma . \tag{32.6}$$

Introducing in this relation the result (32.5) we find:

$$h_{00} = h_{11} = h_{22} = h_{33} = \frac{-\kappa c^2}{4\pi G}\varphi ; \qquad h_{\mu\nu} = 0 \quad \text{for} \quad \mu \neq \nu . \tag{32.7}$$

Therefore, in the linear approximation the metric corresponding to the material distribution (32.1) is:

$$\mathrm{d}s^2 = \left(1 - \frac{\kappa c^2 \varphi}{4\pi G}\right)\mathrm{d}t^2 - \left(1 + \frac{\kappa c^2 \varphi}{4\pi G}\right)(\mathrm{d}x^2 + \mathrm{d}y^2 + \mathrm{d}z^2). \tag{32.8}$$

Note that this formula determines the field inside as well as outside the material distribution.

Let us now consider the case of a spherically symmetric material distribution,

$$\varrho = \varrho(r),$$

where r is the distance from the centre of symmetry. Then the solution of Equation (32.4) outside the matter is:

$$\varphi = \frac{GM}{r}, \qquad M = \int \varrho \, \mathrm{d}^3 x . \tag{32.9}$$

The metric (32.8) with φ given by (32.9) has the form of the linearized isotropic Schwarzschild metric (exercice at the end of Chapter V):

$$\mathrm{d}s^2 = \left(1 - \frac{2m}{r}\right)\mathrm{d}t^2 - \left(1 + \frac{2m}{r}\right)(\mathrm{d}x^2 + \mathrm{d}y^2 + \mathrm{d}z^2). \tag{32.10}$$

In order to make the expressions (32.8) with (32.9) and (32.10) identical we must put:

$$\frac{\kappa c^2}{4\pi G} GM = 2m . \tag{32.11}$$

Since we know the relation (22.18) between m and M we find:

$$\kappa = \frac{8\pi G}{c^4} \tag{32.12}$$

which is the result (19.13) introduced already in Section 19.

The solution of (32.4) inside the body can be found without difficulty. If the density ϱ is constant, one finds that the maximum value of φ corresponds to $r=0$ and is $\frac{3}{2}$ times larger than the value of φ at the surface of the body. In the case of the Sun the density increases appreciably towards the centre and consequently the central value of φ will be larger. According to present estimates the central value of φ would be at most of the order of 10 times the value at the surface: The field will therefore be weak everywhere, as we mentioned in the beginning of Section 31.

One last remark: If the material distribution is not spherically symmetric, we shall have instead of (32.9):

$$\varphi = \frac{GM}{r} + \cdots, \tag{32.13}$$

the dots representing higher multipoles which vanish at large distances at least as $1/r^2$. Thus the coefficient of the term $\sim 1/r$ in g_{00} determines the mass of the body also in this case. This result, derived here from the linearized field equation, is valid generally as a consequence of the conservation laws which we shall establish in Section 35.

A2. Central body with angular momentum

We shall assume that the body rotates around the x^3-axis as an axis of symmetry. Therefore the matter distribution and the gravitational field will be, in spite of the rotation, time-independent or, as we say in this case, *stationary*. The velocity v is assumed to be small compared with c, $v/c \ll 1$. We shall retain only the terms which are linear in v/c. Therefore the matter tensor will now be of the form (Figure 17):

$$\begin{aligned} &{}_0T_{00} = \varrho c^2\,; \qquad {}_0T_{01} = \varrho c v \sin\varphi\,, \qquad {}_0T_{02} = -\,\varrho c v \cos\varphi\,; \\ &\text{all other} \quad {}_0T_{\mu\nu} = 0\,. \end{aligned} \tag{32.14}$$

Because of the axial symmetry the quantities ϱ and v are:

$$\varrho = \varrho(\tilde{R}, x^3)\,, \qquad v = v(\tilde{R}, x^3)\,; \qquad \tilde{R}^2 = (x^1)^2 + (x^2)^2\,. \tag{32.15}$$

It is then verified immediately that the conservation law (31.14) is satisfied.

Since the field is stationary, the linearized field equation (31.13) reduces again to the Poisson equation (32.2). We now have the following three non-trivial equations:

$$\Delta\gamma_{00} = 2\kappa\varrho c^2\,; \qquad \Delta\gamma_{01} = 2\kappa\,{}_0T_{01}\,, \qquad \Delta\gamma_{02} = 2\kappa\,{}_0T_{02}\,. \tag{32.16}$$

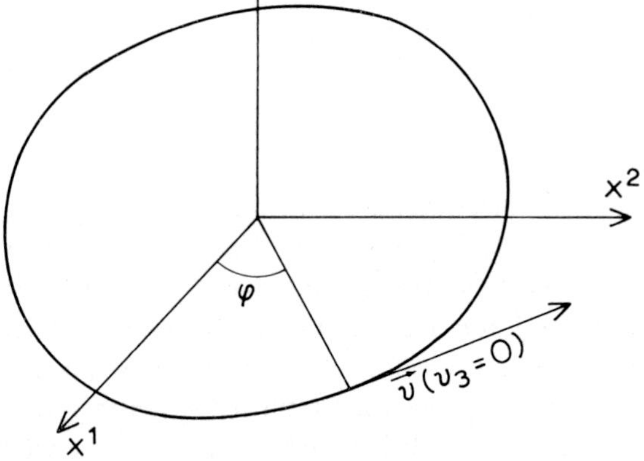

Fig. 17. Axial symmetry: the angle φ and the velocity $\mathbf{v}$.

The remaining $\gamma_{\mu\nu}$ satisfy the equation $\Delta\gamma_{\mu\nu}=0$ everywhere and consequently they must vanish.

We shall determine here the exterior solution only. The first Equation (32.16) is identical with (32.3) and consequently we shall have again the solution (32.5). With (32.13) we find:

$$\gamma_{00} = -\frac{4m}{r} + \cdots; \qquad m = \frac{GM}{c^2}, \qquad M = \int \varrho \, \mathrm{d}^3 x. \tag{32.17}$$

The solutions of the last two Equations (32.16) will be determined from the general formula (31.24):

$$\gamma_{01} = -\frac{\kappa}{2\pi} \int \frac{1}{R} \, {}_0T_{01}(X^i) \, \mathrm{d}^3 X, \qquad \gamma_{02} = -\frac{\kappa}{2\pi} \int \frac{1}{R} \, {}_0T_{02}(X^i) \, \mathrm{d}^3 X. \tag{32.18}$$

The quantities ${}_0T_{01}$ and ${}_0T_{02}$ differ from ${}_0T_{00}$ in one important respect: They change sign when x^2 and x^3 change sign. This means that ${}_0T_{01}$ and ${}_0T_{02}$ have a *dipole structure*. In order to calculate the integrals (32.18) at distances which are large compared with the dimensions of the body we write (31.25) in detail (see Figure 18):

$$R^2 = r^2 - 2x^i X^i + X^i X^i, \qquad r^2 = x^i x^i.$$

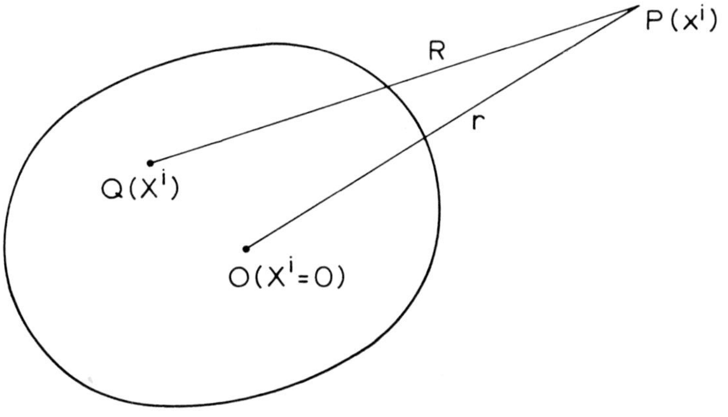

Fig. 18. The distances R and r.

We have omitted the summation symbol Σ_i. At large distances, $X^i/r \ll 1$, we find:

$$\frac{1}{R} = \frac{1}{r}\left(1 - \frac{2x^i X^i}{r^2} + \cdots\right)^{-1/2} = \frac{1}{r} + \frac{x^i X^i}{r^3} + \cdots. \tag{32.19}$$

Introducing the expression (32.19) into the first integral (32.18) we find:

$$\int \frac{1}{R} \, {}_0T_{01} \, \mathrm{d}^3 X = \frac{1}{r} \int {}_0T_{01} \, \mathrm{d}^3 X + \frac{x^i}{r^3} \int X^i \, {}_0T_{01} \, \mathrm{d}^3 X + \cdots.$$

The coefficient of $1/r$ vanishes because of the axial symmetry. Because of the form (32.14) of ${}_0T_{01}$ the second integral is different from zero only for $i=2$. Therefore:

$$\gamma_{01} = \frac{-\kappa x^2}{2\pi r^3} \int X^2 \, {}_0T_{01} \, \mathrm{d}^3X = \frac{\kappa x^2}{2\pi r^3} \int X^2 \, {}_0T^{01} \, \mathrm{d}^3X . \tag{32.20}$$

In a similar way we find:

$$\gamma_{02} = \frac{\kappa x^1}{2\pi r^3} \int X^1 \, {}_0T^{02} \, \mathrm{d}^3X . \tag{32.21}$$

The x^3-component of the angular momentum is given in special relativity by the following relation:

$$J^3 = \frac{1}{c} \int (X^1 \, {}_0T^{02} - X^2 \, {}_0T^{01}) \, \mathrm{d}^3X . \tag{32.22}$$

The following relation can be verified at once as a consequence of the axial symmetry:

$$\int X^1 \, {}_0T^{02} \, \mathrm{d}^3X = - \int X^2 \, {}_0T^{01} \, \mathrm{d}^3X . \tag{32.23}$$

We mention without proof that this relation follows generally for any time-independent material distribution as a consequence of the conservation law (31.14). With (32.23) we find from (32.22):

$$\int X^1 \, {}_0T^{02} \, \mathrm{d}^3X = - \int X^2 \, {}_0T^{01} \, \mathrm{d}^3X = \frac{c}{2} J^3 . \tag{32.24}$$

Introducing these values in (32.20) and (32.21) we find finally:

$$\gamma_{01} = - \frac{\kappa c x^2}{4\pi r^3} J^3 , \qquad \gamma_{02} = \frac{\kappa c x^1}{4\pi r^3} J^3 . \tag{32.25}$$

The non-vanishing quantities $h_{\mu\nu}$ are now the following:

$$h_{00} = h_{11} = h_{22} = h_{33} = \tfrac{1}{2}\gamma_{00} = - \frac{2m}{r} ; \tag{32.26}$$

$$h_{01} = \gamma_{01} = \frac{-2G}{c^3} \frac{x^2}{r^3} J^3 , \qquad h_{02} = \gamma_{02} = \frac{2G}{c^3} \frac{x^1}{r^3} J^3 . \tag{32.27}$$

The interesting information contained in these results is that the angular momentum of the central body manifests itself through special terms appearing in the components g_{0i}. Equation (32.27) shows the structure of these terms in the case $\mathbf{J} \parallel Ox^3$. In the general case,

$$\mathbf{J} = (J^1, J^2, J^3) \tag{32.28}$$

we shall have:

$$g_{0i} = h_{0i} + \cdots = \frac{G}{c^3 r^3} \varepsilon_{ikl} (J^k x^l - J^l x^k) + \cdots , \tag{32.29}$$

ε_{ikl} being the 3-dimensional permutation symbol:

$$\varepsilon_{ikl} = -\,\varepsilon_{kil} = -\,\varepsilon_{ilk}\,; \qquad \varepsilon_{123} = +1\,. \tag{32.30}$$

This result, derived here from the linearized field equation, has a general validity: It can be derived also from the general form of the conservation law of angular momentum which we shall establish in Section 35.

Exercise

VIII: Show that in the case of a ${}_0T^{\mu\nu}$ which is time-independent and bounded in the 3-dimensional space the conservation law

$${}_0T^{\mu\nu}{}_{,\nu} = 0$$

leads directly to the relation:

$$\int\limits_{t=\mathrm{const}} (x^i\,{}_0T^{k\mu} + x^k\,{}_0T^{i\mu})\,\mathrm{d}^3x = 0,$$

which is a generalisation of (32.23).

CHAPTER VIII

VARIATIONAL PRINCIPLE. IDENTITIES, CONSERVATION LAWS

33. Variational Principle

The method of the variational principle has been developed and used systematically in classical mechanics for deriving the equations of motion of a material system with given properties. The basic quantity is the Lagrange function or *Lagrangian* of the system, depending on the generalized coordinates $\xi^a(t)$ describing the system and in the usual, simplest case on the velocities $\dot{\xi}^a \equiv d\xi^a/dt$:

$$L = L(\xi^a, \dot{\xi}^a).$$

We consider a possible trajectory of the system in the space of the ξ^a between the points $P_1(\xi_1^a)$ and $P_2(\xi_2^a)$,

$$\xi_1^a = \xi^a(t_1), \qquad \xi_2^a = \xi^a(t_2),$$

and we form the *action integral*

$$I = \int_{t_1}^{t_2} L\, dt.$$

The equations of motion follow from the requirement

$$\delta I = 0$$

for any variation $\delta\xi^a$ of the trajectory which leaves the endpoints unchanged (Figure 19).

This method has been extended to field theories, permitting the derivation of the field equations of the theory from a variational principle. The Lagrange function will now depend on the quantities φ^a describing the field and the φ^a will be functions of the coordinates x^λ of the space-time: Instead of the velocities $\dot{\xi}^a$ which we had before we shall now have the partial derivatives $\varphi^a{}_{,\lambda} \equiv \partial\varphi^a/\partial x^\lambda$. In the usual simplest case the lagrangian of the field will depend on φ^a and $\varphi^a{}_{,\lambda}$:

$$\mathfrak{L} = \mathfrak{L}(\varphi^a, \varphi^a{}_{,\lambda}). \tag{33.1}$$

Instead of the interval $(t_1 t_2)$ considered in the previous case we have now to consider a given region Ω of the space-time. The action integral will be:

$$I = \int_\Omega \mathfrak{L}\, d\omega, \qquad d\omega = dx^1\, dx^2\, dx^3\, dx^4. \tag{33.2}$$

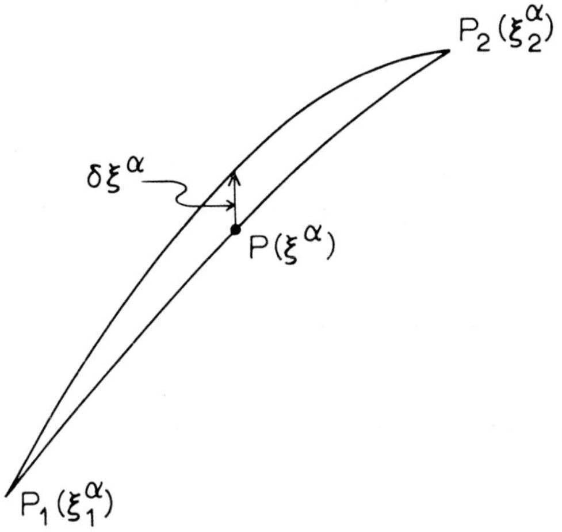

Fig. 19. Variation of a trajectory leaving the endpoints unchanged.

The field equations are then derived from the requirement

$$\delta I = 0\,, \tag{33.3}$$

the allowed variations $\delta\varphi^a$ of φ^a being arbitrary inside Ω, but vanishing on the boundary Σ of Ω:

$$\delta\varphi^a = 0 \quad \text{on} \quad \Sigma\,. \tag{33.4}$$

The quantities φ^a have in general tensor character: In Maxwell theory they are the components of the vector potential A_λ, in general relativity the components of the metric $g_{\lambda\mu}$ etc. The lagrangian $\mathfrak{L}$ has to be a scalar density, in which case the action I will be independent of the coordinate system used in the evaluation of the integral (33.2). The resulting field equations will then be tensor equations, having the same form in all coordinate systems. We express this fact by saying that the field equations are *covariant*.

The calculation needed to derive the field equations is very simple. From (33.2) we find:

$$\delta I = \int_\Omega \delta\mathfrak{L}\,\mathrm{d}\omega\,. \tag{33.5}$$

Since $\mathfrak{L}$ depends on φ^a and $\varphi^a{}_{,\lambda}$, we have the elementary formula:

$$\delta\mathfrak{L} = \frac{\partial\mathfrak{L}}{\partial\varphi^a}\,\delta\varphi^a + \frac{\partial\mathfrak{L}}{\partial\varphi^a{}_{,\lambda}}\,\delta\varphi^a{}_{,\lambda}\,. \tag{33.6}$$

The variations $\delta\varphi^a$ and $\delta\varphi^a{}_{,\lambda}$ are not independent:

$$\delta\varphi^a{}_{,\lambda} = (\delta\varphi^a)_{,\lambda}\,. \tag{33.7}$$

We therefore can write $\delta\mathfrak{L}$ in the form:

$$\partial\mathfrak{L} = \left\{\frac{\partial\mathfrak{L}}{\partial\varphi^a} - \frac{\partial}{\partial x^\lambda}\left(\frac{\partial\mathfrak{L}}{\partial\varphi^a{}_{,\lambda}}\right)\right\}\delta\varphi^a + \frac{\partial}{\partial x^\lambda}\left(\frac{\partial\mathfrak{L}}{\partial\varphi^a{}_{,\lambda}}\,\delta\varphi^a\right). \tag{33.8}$$

When we introduce the expression (33.8) into (33.5) we see that the integral of the last term can be transformed into an integral over Σ, which vanishes because of (33.4). We therefore have:

$$\delta I = \int_\Omega \left\{\frac{\partial\mathfrak{L}}{\partial\varphi^a} - \frac{\partial}{\partial x^\lambda}\left(\frac{\partial\mathfrak{L}}{\partial\varphi^a{}_{,\lambda}}\right)\right\}\delta\varphi^a\, d\omega. \tag{33.9}$$

Since the variations $\delta\varphi^a$ can be chosen arbitrarily inside Ω, we see that the requirement (33.3) will be fulfilled if and only if we have at each point of Ω:

$$\frac{\partial\mathfrak{L}}{\partial\varphi^a} - \frac{\partial}{\partial x^\lambda}\left(\frac{\partial\mathfrak{L}}{\partial\varphi^a{}_{,\lambda}}\right) = 0. \tag{33.10}$$

This is the field equation derived from the variational principle based on the Lagrangian $\mathfrak{L}$.

The more general case of a Lagrangian $\mathfrak{L}$ depending also on the second derivatives $\varphi^a{}_{,\lambda\mu}$ of the field quantities,

$$\mathfrak{L} = \mathfrak{L}(\varphi^a, \varphi^a{}_{,\lambda}, \varphi^a{}_{,\lambda\mu}),$$

can be discussed in a similar way. One has now to demand that not only the variations $\delta\varphi^a$, but also the derivatives $\delta\varphi^a{}_{,\lambda}$ vanish on Σ. A calculation similar to the preceding one leads then immediately to the following field equation:

$$\frac{\partial\mathfrak{L}}{\partial\varphi^a} - \frac{\partial}{\partial x^\lambda}\left(\frac{\partial\mathfrak{L}}{\partial\varphi^a{}_{,\lambda}}\right) + \frac{\partial^2}{\partial x^\lambda \partial x^\mu}\left(\frac{\partial\mathfrak{L}}{\partial\varphi^a{}_{,\lambda\mu}}\right) = 0. \tag{33.11}$$

The field variables of the gravitational field will evidently be the components of the metric, e.g. the covariant ones $g_{\lambda\mu}$. The Lagrangian has to be a scalar density constructed from $g_{\lambda\mu}$ and its first and eventually second derivatives. The simplest scalar density of this type is the one derived from the scalar R, Equation (14.10):

$$\sqrt{-g}\, g^{\lambda\mu} R_{\lambda\mu} \equiv \mathfrak{g}^{\lambda\mu} R_{\lambda\mu} \equiv \mathfrak{R}. \tag{33.12'}$$

This is actually the Lagrangian of the gravitational field in empty space:

$$\mathfrak{L} = \mathfrak{R}. \tag{33.12}$$

The field equations for empty space,

$$R_{\lambda\mu} = 0, \tag{33.13}$$

can be derived from (33.12) by a rather long but otherwise routine calculation using the general result (33.11). One can show that (33.12) is the only scalar Lagrangian which yields second order field equations.

The calculations will be simpler if we start from a reduced Lagrangian obtained in the following way. Using the detailed expression (19.8) for $R_{\lambda\mu}$ we find:

$$\mathfrak{L} = \mathfrak{g}^{\lambda\mu}(-\Gamma^{\alpha}_{\lambda\mu,\alpha} + \Gamma^{\alpha}_{\lambda\alpha,\mu} - \Gamma^{\beta}_{\lambda\mu}\Gamma^{\alpha}_{\beta\alpha} + \Gamma^{\beta}_{\lambda\alpha}\Gamma^{\alpha}_{\beta\mu}). \tag{33.14}$$

We now write:

$$\mathfrak{g}^{\lambda\mu}(-\Gamma^{\alpha}_{\lambda\mu,\alpha} + \Gamma^{\alpha}_{\lambda\alpha,\mu}) = (-\mathfrak{g}^{\lambda\mu}\Gamma^{\alpha}_{\lambda\mu} + \mathfrak{g}^{\lambda\alpha}\Gamma^{\mu}_{\lambda\mu})_{,\alpha} + \mathfrak{g}^{\lambda\mu}{}_{,\alpha}\Gamma^{\alpha}_{\lambda\mu} - \mathfrak{g}^{\lambda\mu}{}_{,\mu}\Gamma^{\alpha}_{\lambda\alpha}. \tag{33.15}$$

From (12.11) and (12.17) we have:

$$\mathfrak{g}^{\lambda\mu}{}_{;\alpha} = 0.$$

From this relation we derive, with the help of (6.20):

$$\mathfrak{g}^{\lambda\mu}{}_{,\alpha} = -\Gamma^{\lambda}_{\beta\alpha}\mathfrak{g}^{\beta\mu} - \Gamma^{\mu}_{\beta\alpha}\mathfrak{g}^{\lambda\beta} + \Gamma^{\beta}_{\alpha\beta}\mathfrak{g}^{\lambda\mu}. \tag{33.16}$$

Putting $\alpha = \mu$ in this last relation we have:

$$\mathfrak{g}^{\lambda\mu}{}_{,\mu} = -\Gamma^{\lambda}_{\beta\mu}\mathfrak{g}^{\beta\mu}. \tag{33.17}$$

Introducing the expressions (33.16) and (33.17) in (33.15) we find:

$$\mathfrak{g}^{\lambda\mu}(-\Gamma^{\alpha}_{\lambda\mu,\alpha} + \Gamma^{\alpha}_{\lambda\alpha,\mu}) = (-\mathfrak{g}^{\lambda\mu}\Gamma^{\alpha}_{\lambda\mu} + \mathfrak{g}^{\lambda\alpha}\Gamma^{\mu}_{\lambda\mu})_{,\alpha} + 2\mathfrak{g}^{\lambda\mu}(\Gamma^{\alpha}_{\lambda\mu}\Gamma^{\beta}_{\alpha\beta} - \Gamma^{\alpha}_{\mu\beta}\Gamma^{\beta}_{\lambda\alpha}).$$

Thus we find finally:

$$\mathfrak{L} = \mathfrak{A}^{\alpha}{}_{,\alpha} + \mathfrak{L}'; \tag{33.18}$$

$$\mathfrak{A}^{\alpha} = -\mathfrak{g}^{\lambda\mu}\Gamma^{\alpha}_{\lambda\mu} + \mathfrak{g}^{\lambda\alpha}\Gamma^{\mu}_{\lambda\mu}, \tag{33.19}$$

$$\mathfrak{L}' = \mathfrak{g}^{\lambda\mu}(\Gamma^{\alpha}_{\lambda\mu}\Gamma^{\beta}_{\alpha\beta} - \Gamma^{\alpha}_{\mu\beta}\Gamma^{\beta}_{\lambda\alpha}). \tag{33.20}$$

One can now prove that a term of the form $\mathfrak{A}^{\alpha}{}_{,\alpha}$ in the lagrangian does not contribute to the field equations and can therefore be neglected. Indeed, this term gives to the integral (33.2) the contribution

$$\int_{\Omega} \mathfrak{A}^{\alpha}{}_{,\alpha}\, d\omega = \int_{\Sigma} \mathfrak{A}^{\alpha}\, d\sigma_{\alpha},$$

which has a vanishing variation because the variations of the field quantities vanish on Σ. (Note, however, that since $\mathfrak{A}^{\alpha}$ contains also derivatives of $g_{\lambda\mu}$ we have to demand now the vanishing on Σ of $\delta g_{\lambda\mu}$ and of $\delta g_{\lambda\mu,\nu}$.) We can therefore derive the field equations from the reduced Lagrangian $\mathfrak{L}'$.

Since $\mathfrak{L}'$ contains only first derivatives of $g_{\lambda\mu}$, it is now sufficient to use the Equation (33.10). We give the detailed results of the calculations for the case in which we take as field variables the components of the contravariant density $\mathfrak{g}^{\lambda\mu}$:

$$\frac{\partial \mathfrak{L}'}{\partial \mathfrak{g}^{\lambda\mu}} \equiv \mathfrak{L}'_{\lambda\mu} = \Gamma^{\alpha}_{\lambda\beta}\Gamma^{\beta}_{\alpha\mu} - \Gamma^{\alpha}_{\lambda\mu}\Gamma^{\beta}_{\alpha\beta}, \tag{33.21}$$

$$\frac{\partial \mathfrak{L}'}{\partial \mathfrak{g}^{\lambda\mu}{}_{,\nu}} \equiv \mathfrak{L}'^{\nu}_{\lambda\mu} = \Gamma^{\nu}_{\lambda\mu} - \tfrac{1}{2}(\delta^{\nu}_{\lambda}\Gamma^{\alpha}_{\alpha\mu} + \delta^{\nu}_{\mu}\Gamma^{\alpha}_{\alpha\lambda}); \tag{33.22}$$

$$\mathfrak{L}'_{\lambda\mu} - \mathfrak{L}'^{\nu}_{\lambda\mu,\,\nu} = R_{\lambda\mu}\,. \tag{33.23}$$

The last relation shows that the vacuum field equation has the form (33.13).

The field equation (33.13) can be derived also from the Lagrangian (33.12) in a direct and elegant way, which we shall describe briefly here. We have:

$$\delta\mathfrak{L} = \mathfrak{g}^{\lambda\mu}\delta R_{\lambda\mu} + \delta\mathfrak{g}^{\lambda\mu}\cdot R_{\lambda\mu}\,. \tag{33.24}$$

From the expression (19.8) for $R_{\lambda\mu}$ we find:

$$\delta R_{\lambda\mu} = -\,(\delta\Gamma^{\alpha}_{\lambda\mu})_{,\,\alpha} + (\delta\Gamma^{\alpha}_{\lambda\alpha})_{,\,\mu} - \delta\Gamma^{\beta}_{\lambda\mu}\cdot\Gamma^{\alpha}_{\beta\alpha} - \Gamma^{\beta}_{\lambda\mu}\delta\Gamma^{\alpha}_{\beta\alpha} + \\ + \delta\Gamma^{\beta}_{\lambda\alpha}\cdot\Gamma^{\alpha}_{\beta\mu} + \Gamma^{\beta}_{\lambda\alpha}\delta\Gamma^{\alpha}_{\beta\mu}\,. \tag{33.25}$$

The variation $\delta\Gamma^{\alpha}_{\lambda\mu}$ is the difference of two connections and therefore according to (5.6) a tensor. We verify at once that the relation (33.25) is the detailed form of the *Palatini equation*:

$$\delta R_{\lambda\mu} = -\,(\delta\Gamma^{\alpha}_{\lambda\mu})_{;\,\alpha} + (\delta\Gamma^{\alpha}_{\lambda\alpha})_{;\,\mu}\,. \tag{33.26}$$

We then find:

$$\mathfrak{g}^{\lambda\mu}\delta R_{\lambda\mu} = (-\,\mathfrak{g}^{\lambda\mu}\delta\Gamma^{\alpha}_{\lambda\mu} + \mathfrak{g}^{\lambda\alpha}\delta\Gamma^{\mu}_{\lambda\mu})_{;\,\alpha}\,. \tag{33.27}$$

The quantity in the bracket is a contravariant vector density and consequently the covariant derivative can be replaced, according to (6.19), by the ordinary derivative. Hence:

$$\delta\mathfrak{L} = (-\,\mathfrak{g}^{\lambda\mu}\delta\Gamma^{\alpha}_{\lambda\mu} + \mathfrak{g}^{\lambda\alpha}\delta\Gamma^{\mu}_{\lambda\mu})_{,\,\alpha} + R_{\lambda\mu}\delta\mathfrak{g}^{\lambda\mu}\,. \tag{33.28}$$

The first term of $\delta\mathfrak{L}$ is of the form $\mathfrak{B}^{\alpha}{}_{,\alpha}$ and therefore it gives no contribution to δI. We thus have finally:

$$\delta I = \int_{\Omega} R_{\lambda\mu}\delta\mathfrak{g}^{\lambda\mu}\,\mathrm{d}\omega\,,$$

leading again to the field equation (33.13).

The field equation in the presence of matter, $T_{\mu\nu} \neq 0$, can be derived formally from the Lagrangian:

$$\mathfrak{L} = g^{\lambda\mu}(\sqrt{-g}\;R_{\lambda\mu} + \kappa\mathfrak{T}_{\lambda\mu})\,, \qquad \mathfrak{T}_{\lambda\mu} \equiv \sqrt{-g}\;T_{\lambda\mu}\,. \tag{33.29}$$

We have to use now as field variables the $g^{\lambda\mu}$ and to allow no variation of the quantities $\mathfrak{T}_{\lambda\mu}$. Taking into account the formula

$$\delta\sqrt{-g} = -\,\tfrac{1}{2}\sqrt{-g}\;g_{\lambda\mu}\delta g^{\lambda\mu}\,, \tag{33.30}$$

which can be derived easily from (6.10), we find from (33.29):

$$\delta\mathfrak{L} = \delta g^{\lambda\mu}(\sqrt{-g}R_{\lambda\mu} + \kappa\mathfrak{T}_{\lambda\mu}) + g^{\lambda\mu}\sqrt{-g}\;\delta R_{\lambda\mu} - \tfrac{1}{2}\sqrt{-g}\;g_{\lambda\mu}\delta g^{\lambda\mu}\cdot R\,.$$

The second term on the right-hand side has, according to (33.27), the form $\mathfrak{B}^{\alpha}{}_{,\alpha}$ and can be omitted. The remaining terms lead to the field equation

$$\sqrt{-g}\,(R_{\lambda\mu} - \tfrac{1}{2}g_{\lambda\mu}R) = -\,\kappa\mathfrak{T}_{\lambda\mu}, \tag{33.31}$$

which is equivalent to (19.7).

34. Identities Corresponding to the Lagrangian $\mathfrak{L}'$

According to the theorem of Noether there is a system of identities corresponding to a given Lagrangian $\mathfrak{L}$ which is a scalar density. We shall develop the general method for deriving these identities and then we shall apply it in order to derive the identities corresponding to the gravitational Lagrangian $\mathfrak{L}'$ given by Equation (33.20). We are interested in these identities because they lead to the physical conservation laws, as we shall show in Section 35. When we are dealing with the Lagrangian $\mathfrak{L}'$ we have to remember the following two remarks:

(1) $\mathfrak{L}'$ is not a true scalar density; it transforms as a scalar density only for linear transformations.

(2) $\mathfrak{L}'$ differs from the true scalar density $\mathfrak{L}$, Equation (33.12), by a term having the form of a divergence, as shown by Equation (33.18).

The basic remark which leads to these identities in the case of a true scalar density $\mathfrak{L}$ is that, when the field variables and the integration domain Ω are given, the value of the action integral I will be the same if we calculate it in any coordinate system. We shall consider two coordinate systems x^{μ} and $\tilde{x}^{\mu}$ related by the 'infinitesimal' transformation ($\varepsilon \to 0$):

$$\tilde{x}^{\mu} = x^{\mu} + \varepsilon\xi^{\mu}(x^{\nu}). \tag{34.1}$$

In the case of a true scalar density $\mathfrak{L}$ we shall have

$$\tilde{I} = I; \qquad \tilde{I} = \int_{\Omega} \tilde{\mathfrak{L}}\,\mathrm{d}\tilde{\omega}, \qquad I = \int_{\Omega} \mathfrak{L}\,\mathrm{d}\omega, \tag{34.2}$$

for any functions $\xi^{\mu}(x^{\nu})$.

In the case of the Lagrangian $\mathfrak{L}'$ we shall have

$$\tilde{I}' = I'; \qquad \tilde{I}' = \int_{\Omega} \tilde{\mathfrak{L}}'\,\mathrm{d}\tilde{\omega}, \qquad I' = \int_{\Omega} \mathfrak{L}'\,\mathrm{d}\omega, \tag{34.3}$$

only in the following two cases:

(1) The ξ^{μ} are linear functions of x^{ν}, in which case the transformation (34.1) is linear.

(2) The ξ^{ν} are arbitrary inside Ω but they vanish on Σ with their first and second derivatives. Indeed, it follows from (33.18) and (34.3):

$$\tilde{I}' - I' = \int_{\Sigma} (\mathfrak{A}^{\alpha} - \tilde{\mathfrak{A}}^{\alpha})\,\mathrm{d}\sigma_{\alpha}$$

and the difference $\mathfrak{A}^\alpha - \tilde{\mathfrak{A}}^\alpha$ vanishes on Σ. The vanishing on Σ of the derivatives $\xi^\mu{}_{,\lambda}$ and $\xi^\mu{}_{,\lambda\nu}$ is necessary because $\mathfrak{A}^\alpha$ contains $\mathfrak{g}^{\lambda\mu}$ and $\mathfrak{g}^{\lambda\mu}{}_{,\nu}$. (See Equations (34.10) and (34.11).)

We shall now determine a detailed expression for the difference $\tilde{I}' - I'$. It will be convenient to consider the transformation (34.1) as a mapping of the space onto itself. The quantity $\tilde{I}'$ will then be an integral over the region $\tilde{\Omega}$ having the new boundary $\tilde{\Sigma}$ (Figure 20). Therefore:

$$\tilde{I}' = \int_{\Omega} \tilde{\mathfrak{L}}' \, \mathrm{d}\omega + \int_{\tilde{\Omega}-\Omega} \tilde{\mathfrak{L}}' \, \mathrm{d}\omega , \tag{34.4}$$

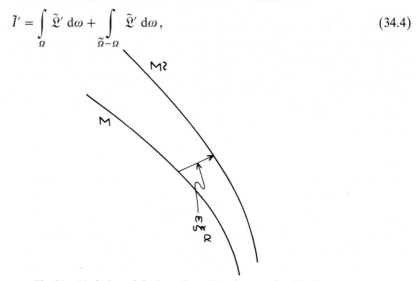

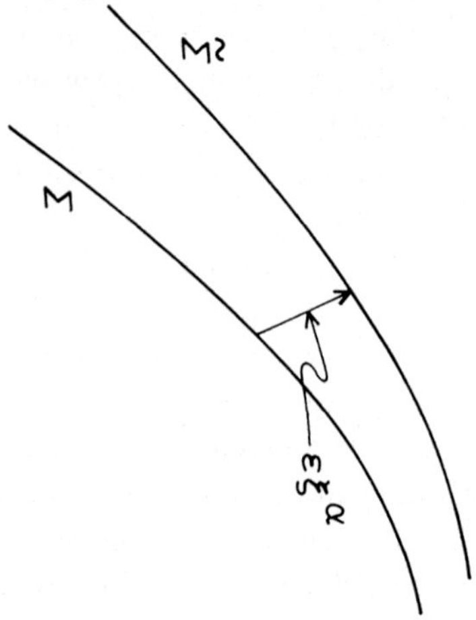

Fig. 20. Variation of the boundary Σ by the mapping (34.1).

$\tilde{\Omega} - \Omega$ being the region between $\tilde{\Sigma}$ and Σ. The second integral can be written as an integral over Σ and then again as an integral over Ω:

$$\int_{\tilde{\Omega}-\Omega} \tilde{\mathfrak{L}}' \, \mathrm{d}\omega = \int_{\Sigma} \tilde{\mathfrak{L}}' \varepsilon \xi^\mu \, \mathrm{d}\sigma_\mu = \varepsilon \int_{\Omega} (\mathfrak{L}' \xi^\mu)_{,\mu} \, \mathrm{d}\omega .$$

In the last integral we have written $\mathfrak{L}'$ instead of $\tilde{\mathfrak{L}}'$ because we already have the factor ε and we are retaining only terms of the first order. So we have finally:

$$\tilde{I}' - I' = \int_{\Omega} \{\tilde{\mathfrak{L}}' - \mathfrak{L}' + \varepsilon (\mathfrak{L}' \xi^\mu)_{,\mu}\} \, \mathrm{d}\omega . \tag{34.5}$$

We now calculate the difference $\tilde{\mathfrak{L}}' - \mathfrak{L}'$. Remembering that $\mathfrak{L}'$ is a function of $\mathfrak{g}^{\lambda\mu}$ and $\mathfrak{g}^{\lambda\mu}{}_{,\nu}$, we see that we have to start by calculating the difference:

$$\tilde{\mathfrak{g}}^{\lambda\mu}(x^\alpha) - \mathfrak{g}^{\lambda\mu}(x^\alpha) \equiv \delta \mathfrak{g}^{\lambda\mu} . \tag{34.6}$$

The transformation formula for $\mathfrak{g}^{\lambda\mu}$ is, from (1.15) and (3.3):

$$\tilde{\mathfrak{g}}^{\lambda\mu}(\tilde{x}^\alpha) = \frac{\partial \tilde{x}^\lambda}{\partial x^\varrho} \frac{\partial \tilde{x}^\mu}{\partial x^\sigma} \cdot \det \frac{\partial x}{\partial \tilde{x}} \cdot \mathfrak{g}^{\varrho\sigma}(x^\alpha) . \tag{34.7}$$

From (34.1) we derive the result:

$$\det \frac{\partial x}{\partial \tilde{x}} = 1 - \varepsilon \xi^{\varrho}{}_{,\varrho} + \cdots. \tag{34.8}$$

We then find, retaining only terms of first order in ε:

$$\tilde{\mathfrak{g}}^{\lambda\mu}(\tilde{x}^{\alpha}) = \mathfrak{g}^{\lambda\mu}(x^{\alpha}) + \varepsilon(\xi^{\lambda}{}_{,\varrho}\mathfrak{g}^{\varrho\mu} + \xi^{\mu}{}_{,\varrho}\mathfrak{g}^{\lambda\varrho} - \xi^{\varrho}{}_{,\varrho}\mathfrak{g}^{\lambda\mu}). \tag{34.9}$$

In (34.6) we need the quantity $\tilde{\mathfrak{g}}^{\lambda\mu}(x^{\alpha})$. Using again (34.1) we find:

$$\tilde{\mathfrak{g}}^{\lambda\mu}(\tilde{x}^{\alpha}) = \tilde{\mathfrak{g}}^{\lambda\mu}(x^{\alpha}) + \varepsilon \tilde{\mathfrak{g}}^{\lambda\mu}{}_{,\varrho}\xi^{\varrho}.$$

Introducing this result in (34.9) we find:

$$\delta\mathfrak{g}^{\lambda\mu} = \varepsilon(\xi^{\lambda}{}_{,\varrho}\mathfrak{g}^{\varrho\mu} + \xi^{\mu}{}_{,\varrho}\mathfrak{g}^{\lambda\varrho} - \xi^{\varrho}{}_{,\varrho}\mathfrak{g}^{\lambda\mu} - \xi^{\varrho}\mathfrak{g}^{\lambda\mu}{}_{,\varrho}). \tag{34.10}$$

A similar calculation for the derivatives $\mathfrak{g}^{\lambda\mu}{}_{,\nu}$ shows that the variation induced by the transformation (34.1) is given by the simple formula:

$$\tilde{\mathfrak{g}}^{\lambda\mu}{}_{,\nu}(x^{\alpha}) - \mathfrak{g}^{\lambda\mu}{}_{,\nu}(x^{\alpha}) \equiv \delta\mathfrak{g}^{\lambda\mu}{}_{,\nu} = (\delta\mathfrak{g}^{\lambda\mu})_{,\nu}. \tag{34.11}$$

For the variation

$$\tilde{\mathfrak{L}}'(x^{\alpha}) - \mathfrak{L}'(x^{\alpha}) \equiv \delta\mathfrak{L}' \tag{34.12}$$

we have the formula:

$$\delta\mathfrak{L}' = \mathfrak{L}'_{\lambda\mu}\delta\mathfrak{g}^{\lambda\mu} + \mathfrak{L}'^{\nu}_{\lambda\mu}\delta\mathfrak{g}^{\lambda\mu}{}_{,\nu}. \tag{34.13}$$

Taking into account the relation (33.23) we rewrite this equation in the form:

$$\delta\mathfrak{L}' = R_{\lambda\mu}\delta\mathfrak{g}^{\lambda\mu} + (\mathfrak{L}'^{\nu}_{\lambda\mu}\delta\mathfrak{g}^{\lambda\mu})_{,\nu}. \tag{34.14}$$

Introducing the expression (34.10) in (34.14) we find after some rearrangements of terms:

$$\left.\begin{aligned}\delta\mathfrak{L}' = \varepsilon\{&2(R_{\lambda\mu}\mathfrak{g}^{\varrho\lambda}\xi^{\mu})_{,\varrho} - 2(R_{\lambda\mu}\mathfrak{g}^{\varrho\mu})_{,\varrho}\xi^{\lambda} - (\mathfrak{R}\xi^{\varrho})_{,\varrho} + \\ &+ \mathfrak{R}_{,\varrho}\xi^{\varrho} - R_{\lambda\mu}\mathfrak{g}^{\lambda\mu}{}_{,\varrho}\xi^{\varrho}\} + (\mathfrak{L}'^{\nu}_{\lambda\mu}\delta\mathfrak{g}^{\lambda\mu})_{,\nu}.\end{aligned}\right\} \tag{34.15}$$

With this expression for $\delta\mathfrak{L}'$ we may rewrite (34.5) as:

$$\left.\begin{aligned}\tilde{I}' - I' = \varepsilon &\int_{\Omega} \{-2\mathfrak{R}^{\varrho}_{\lambda,\varrho} + \mathfrak{R}_{,\lambda} - R_{\varrho\mu}\mathfrak{g}^{\varrho\mu}{}_{,\lambda}\}\,\xi^{\lambda}\,d\omega + \\ &+ \int_{\Omega} \frac{\partial}{\partial x^{\mu}}\{2\varepsilon R_{\lambda\varrho}\mathfrak{g}^{\mu\varrho}\xi^{\lambda} - \varepsilon\mathfrak{R}\xi^{\mu} + \mathfrak{L}'^{\mu}_{\lambda\varrho}\delta\mathfrak{g}^{\lambda\varrho} + \varepsilon\mathfrak{L}'\xi^{\mu}\}\,d\omega.\end{aligned}\right\} \tag{34.16}$$

From (34.16) we shall now derive 3 identities. The first identity follows from the remark that the left-hand side of (34.16) vanishes when ξ^{μ} is arbitrary in Ω but vanishes on Σ. The last integral in (34.16) vanishes in this case. Since the first integral must now vanish for arbitrary ξ^{μ}, we shall have the relation:

$$2\mathfrak{R}_{\lambda}{}^{\varrho}{}_{,\varrho} + R_{\varrho\mu}\mathfrak{g}^{\varrho\mu}{}_{,\lambda} - \mathfrak{R}_{,\lambda} = 0. \tag{34.17}$$

This relation can be written also in the form:

$$(\mathfrak{R}_\lambda{}^\varrho - \tfrac{1}{2}\delta_\lambda^\varrho\mathfrak{R})_{,\varrho} + \tfrac{1}{2}R_{\varrho\mu}\mathfrak{g}^{\varrho\mu}{}_{,\lambda} = 0$$

and is then seen to be equivalent to the already known identity (19.6):

$$(\mathfrak{R}_\lambda{}^\varrho - \tfrac{1}{2}\delta_\lambda^\varrho\mathfrak{R})_{;\varrho} = 0 . \tag{34.18}$$

With the result (34.17) the right-hand side of (34.16) is reduced to its second term. Introducing into it the expression (34.10) for $\delta\mathfrak{g}^{\lambda\mu}$ we find:

$$\left.\begin{aligned} \tilde{I}' - I' = \varepsilon \int_\Omega \frac{\partial}{\partial x^\mu} \{(2\mathfrak{R}_\lambda{}^\mu - \mathfrak{R}\delta_\lambda^\mu + \mathfrak{L}'\delta_\lambda^\mu - \mathfrak{L}'^\mu_{\varrho\sigma}\mathfrak{g}^{\varrho\sigma}{}_{,\lambda})\,\xi^\lambda + \\ + \mathfrak{L}'^\mu_{\varrho\sigma}(2\xi^\varrho{}_{,\alpha}\mathfrak{g}^{\alpha\sigma} - \xi^\alpha{}_{,\alpha}\mathfrak{g}^{\varrho\sigma})\}\, d\omega . \end{aligned}\right\} \tag{34.19}$$

The quantity (34.19) vanishes for all ξ^μ which are linear functions of x^α. The simplest ξ^μ of this type are those describing the translations,

$$\xi^\mu = \text{arbitrary constants.}$$

For this case we find from (34.19):

$$\{2\mathfrak{R}_\lambda{}^\mu - \mathfrak{R}\delta_\lambda^\mu + \mathfrak{L}'\delta_\lambda^\mu - \mathfrak{L}'^\mu_{\varrho\sigma}\mathfrak{g}^{\varrho\sigma}{}_{,\lambda}\}_{,\mu} = 0 . \tag{34.20}$$

This is the second identity.

With (34.20) the relation (34.19) is simplified to the form:

$$\left.\begin{aligned} \tilde{I}' - I' = \varepsilon \int_\Omega \{(2\mathfrak{R}_\lambda{}^\mu - \mathfrak{R}\delta_\lambda^\mu + \mathfrak{L}'\delta_\lambda^\mu - \mathfrak{L}'^\mu_{\varrho\sigma}\mathfrak{g}^{\varrho\sigma}{}_{,\lambda})\,\xi^\lambda{}_{,\mu} + \\ + \frac{\partial}{\partial x^\mu}(\mathfrak{L}'^\mu_{\varrho\sigma}\{2\xi^\varrho{}_{,\alpha}\mathfrak{g}^{\alpha\sigma} - \xi^\alpha{}_{,\alpha}\mathfrak{g}^{\varrho\sigma}\})\}\, d\omega . \end{aligned}\right\} \tag{34.21}$$

We now consider the more general form of linear transformations:

$$\xi^\lambda{}_{,\mu} = \text{arbitrary constants.}$$

This gives us the last identity which can be derived from $\mathfrak{L}'$. It is the following:

$$2\mathfrak{R}_\lambda{}^\mu - \mathfrak{R}\delta_\lambda^\mu + \mathfrak{L}'\delta_\lambda^\mu - \mathfrak{L}'^\mu_{\varrho\sigma}\mathfrak{g}^{\varrho\sigma}{}_{,\lambda} = (\delta_\lambda^\mu\mathfrak{L}'^\alpha_{\varrho\sigma}\mathfrak{g}^{\varrho\sigma} - 2\mathfrak{L}'^\alpha_{\lambda\sigma}\mathfrak{g}^{\mu\sigma})_{,\alpha} . \tag{34.22}$$

Using the field equation (33.31) we write the first identity (34.18) in the form:

$$\mathfrak{T}_\lambda{}^\varrho{}_{;\varrho} = 0 . \tag{34.23}$$

The second identity (34.20) can then be written as follows:

$$(\mathfrak{T}_\lambda{}^\varrho + \mathfrak{t}_\lambda{}^\varrho)_{,\varrho} = 0 ; \tag{34.24}$$

$$\kappa\mathfrak{t}_\lambda^\varrho \equiv \tfrac{1}{2}\mathfrak{g}^{\alpha\beta}{}_{,\lambda}\mathfrak{L}'^\varrho_{\alpha\beta} - \tfrac{1}{2}\delta_\lambda^\varrho\mathfrak{L}' . \tag{34.25}$$

The physical meaning of $\mathfrak{t}_\lambda{}^\varrho$ as well as of the identity (34.24) will be discussed in the next section.

The third identity (34.22) allows us to write the sum $\mathfrak{T}_\lambda{}^\varrho + \mathfrak{t}_\lambda{}^\varrho$ in the form of a divergence:

$$\kappa(\mathfrak{T}_\lambda{}^\varrho + \mathfrak{t}_\lambda{}^\varrho) = U_\lambda{}^{\varrho\sigma}{}_{,\sigma}\,; \tag{34.26}$$

$$U_\lambda{}^{\varrho\sigma} = \mathfrak{L}'^{\sigma}_{\lambda\nu}\mathfrak{g}^{\varrho\nu} - \tfrac{1}{2}\delta^\varrho_\lambda \mathfrak{L}'^{\sigma}_{\mu\nu}\mathfrak{g}^{\mu\nu}\,. \tag{34.27}$$

It has now become customary to call the quantity $U_\lambda{}^{\varrho\sigma}$ the *superpotential*.

The detailed expression for the superpotential $U_\lambda{}^{\mu\nu}$ follows from the relations (33.22). If we express the Christoffel symbols in terms of derivatives of $\mathfrak{g}^{\lambda\mu}$, we find the following result:

$$U_\lambda{}^{\mu\nu} = \frac{1}{2\sqrt{-g}}\, g_{\lambda\alpha}(\mathfrak{g}^{\mu\alpha}\mathfrak{g}^{\nu\beta} - \mathfrak{g}^{\nu\alpha}\mathfrak{g}^{\mu\beta})_{,\beta} + \tfrac{1}{2}\delta^\nu_\lambda \mathfrak{g}^{\mu\alpha}{}_{,\alpha} - \tfrac{1}{2}\mathfrak{g}^{\mu\nu}{}_{,\lambda}\,. \tag{34.28}$$

One verifies at once that, when the expression (34.28) is introduced in (34.26), the last two terms of (34.28) do not contribute:

$$(\delta^\nu_\lambda \mathfrak{g}^{\mu\alpha}{}_{,\alpha} - \mathfrak{g}^{\mu\nu}{}_{,\lambda})_{,\nu} = \mathfrak{g}^{\mu\alpha}{}_{,\alpha\lambda} - \mathfrak{g}^{\mu\nu}{}_{,\lambda\nu} = 0\,.$$

Consequently we can use in (34.26) the simpler superpotential

$${}_{\mathrm{F}}U_\lambda{}^{\mu\nu} = \frac{1}{2\sqrt{-g}}\, g_{\lambda\alpha}(\mathfrak{g}^{\mu\alpha}\mathfrak{g}^{\nu\beta} - \mathfrak{g}^{\nu\alpha}\mathfrak{g}^{\mu\beta})_{,\beta}\,. \tag{34.29}$$

This is the *Freud* superpotential, which is antisymmetric in μ, ν:

$${}_{\mathrm{F}}U_\lambda{}^{\mu\nu} = -\,{}_{\mathrm{F}}U_\lambda{}^{\nu\mu}\,. \tag{34.30}$$

We mention here without a proof that ${}_{\mathrm{F}}U_\lambda{}^{\mu\nu}$ is the expression which would appear in (34.26) if we had started from the Lagrangian $\mathfrak{L}$, Equation (33.12), instead of $\mathfrak{L}'$. With (34.28) replaced by (34.29) we have instead of (34.26):

$$\kappa(\mathfrak{T}_\lambda{}^\varrho + \mathfrak{t}_\lambda{}^\varrho) = {}_{\mathrm{F}}U_\lambda{}^{\varrho\sigma}{}_{,\sigma}\,. \tag{34.31}$$

The relation (34.24) follows now at once from the antisymmetry expressed by (34.30).

The identities which we have derived can be put in a variety of different forms by a procedure demonstrated in the following example. We multiply Equation (34.31) by $\mathfrak{g}^{\lambda\mu}$. Taking into account the elementary relation

$$\mathfrak{g}^{\lambda\mu}\,{}_{\mathrm{F}}U_\lambda{}^{\varrho\sigma}{}_{,\sigma} = (\mathfrak{g}^{\lambda\mu}\,{}_{\mathrm{F}}U_\lambda{}^{\varrho\sigma})_{,\sigma} - \mathfrak{g}^{\lambda\mu}{}_{,\sigma}\,{}_{\mathrm{F}}U_\lambda{}^{\varrho\sigma}$$

we arrive at the following new identities:

$$\kappa\sqrt{-g}\,(\mathfrak{T}^{\mu\varrho} + \tilde{\mathfrak{t}}^{\mu\varrho}) = \tilde{U}^{\mu\varrho\sigma}{}_{,\sigma}\,; \tag{34.32}$$

$$\kappa\sqrt{-g}\,\tilde{\mathfrak{t}}^{\mu\varrho} = \kappa\mathfrak{g}^{\lambda\mu}\mathfrak{t}^\varrho_\lambda + \mathfrak{g}^{\lambda\mu}{}_{,\sigma}\,{}_{\mathrm{F}}U_\lambda{}^{\varrho\sigma}\,; \tag{34.33}$$

$$\tilde{U}^{\mu\varrho\sigma} = \mathfrak{g}^{\lambda\mu}\,{}_{\mathrm{F}}U_\lambda{}^{\varrho\sigma} = -\,\tilde{U}^{\mu\sigma\varrho}\,. \tag{34.34}$$

This is the *Landau-Lifshitz* form of the identities.

From (34.29) we find at once:

$$\tilde{U}^{\mu\varrho\sigma} = \tfrac{1}{2}(\mathfrak{g}^{\varrho\mu}\mathfrak{g}^{\sigma\beta} - \mathfrak{g}^{\sigma\mu}\mathfrak{g}^{\varrho\beta})_{,\beta}\,. \tag{34.35}$$

The relation (34.32) takes then the form:

$$\kappa\sqrt{-g}\,(\mathfrak{T}^{\mu\varrho} + \tilde{\mathfrak{t}}^{\mu\varrho}) = \tfrac{1}{2}(\mathfrak{g}^{\varrho\mu}\mathfrak{g}^{\sigma\beta} - \mathfrak{g}^{\sigma\mu}\mathfrak{g}^{\varrho\beta})_{,\beta\sigma}. \tag{34.36}$$

The right-hand side of this relation is symmetric in μ, ϱ. Since $\mathfrak{T}^{\mu\varrho}$ is also symmetric,

$$\mathfrak{T}^{\mu\varrho} = g^{\lambda\mu}\mathfrak{T}_{\lambda}^{\ \varrho} = \sqrt{-g}\,T^{\mu\varrho},$$

it follows that $\tilde{\mathfrak{t}}^{\mu\varrho}$ is also symmetric:

$$\tilde{\mathfrak{t}}^{\mu\varrho} = \tilde{\mathfrak{t}}^{\varrho\mu}. \tag{34.37}$$

The relation

$$\{\sqrt{-g}\,(\mathfrak{T}^{\mu\varrho} + \tilde{\mathfrak{t}}^{\mu\varrho})\}_{,\varrho} = 0 \tag{34.38}$$

follows immediately from the antisymmetry of $\tilde{U}^{\mu\varrho\sigma}$ in ϱ, σ.

35. Conservation Laws

In special relativity, when we use an inertial coordinate system in which the Minkowski metric $\eta_{\mu\nu}$ has the form (31.1), the law of conservation of energy and momentum has the *differential* form:

$$\eta^{\mu\nu}T_{\lambda\mu,\nu} \equiv T_{\lambda\ ,\nu}^{\ \nu} = 0. \tag{35.1}$$

Assuming that we are dealing with a material system of finite 3-dimensional extension (Figure 21), we obtain the *integral* form of the conservation law by integrating Equation (35.1) over a spacelike hypersurface, e.g. the hypersurface t=const:

$$\int\limits_{t=\text{const}} T_{\lambda,\nu}^{\nu}\,\mathrm{d}^3x = \int\limits_{t=\text{const}} T_{\lambda,i}^{i}\,\mathrm{d}^3x + \frac{\mathrm{d}}{\mathrm{d}x^0}\int\limits_{t=\text{const}} T_{\lambda}^{0}\,\mathrm{d}^3x \qquad (i = 1, 2, 3).$$

Since

$$\int\limits_{t=\text{const}} T_{\lambda,i}^{i}\,\mathrm{d}^3x = \int\limits_{S\to\infty} T_{\lambda}^{i}n_i\,\mathrm{d}S = 0,$$

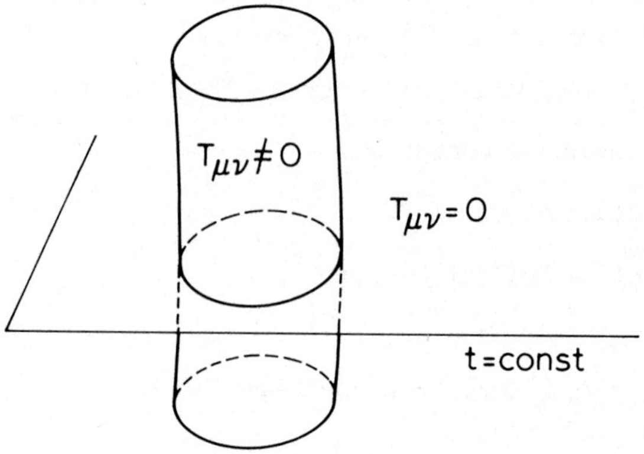

Fig. 21. A material distribution of finite 3-dimensional extension.

because T_λ^i vanishes at large distances, it follows:

$$\frac{\mathrm{d}}{\mathrm{d}x^0} P_\lambda = 0\,, \qquad P_\lambda \equiv \int\limits_{t=\mathrm{const}} T_\lambda^0 \,\mathrm{d}^3x\,. \tag{35.2}$$

The constant vector P_λ is the energy-momentum vector of the material system.

In general relativity the differential form of this conservation law is the covariant generalization of Equation (35.1):

$$\mathfrak{T}_\lambda{}^\nu{}_{;\nu} = 0\,. \tag{35.3}$$

This is identical to Equation (34.23), which in general relativity has been shown to be a consequence of the field equations (33.31).

The detailed form of (35.3) is, according to (6.20):

$$\mathfrak{T}_\lambda{}^\nu{}_{,\nu} - \mathfrak{T}_\varrho{}^\nu \Gamma^\varrho_{\lambda\nu} = 0\,. \tag{35.4}$$

The presence of the second term has the consequence that we cannot derive directly from (35.4) an integral conservation law for the total energy and momentum of the system. This will be possible only if we succeed to write this term in the form of a divergence:

$$-\mathfrak{T}_\varrho{}^\nu \Gamma^\varrho_{\lambda\nu} = \mathfrak{t}_\lambda{}^\nu{}_{,\nu}\,. \tag{35.4$'$}$$

The differential conservation law will then be:

$$(\mathfrak{T}_\lambda{}^\nu + \mathfrak{t}_\lambda{}^\nu)_{,\nu} = 0\,. \tag{35.5}$$

This equation will lead, under certain conditions which we shall examine later, to an integral conservation law of the form (35.2):

$$P_\lambda \equiv \int (\mathfrak{T}_\lambda^0 + \mathfrak{t}_\lambda^0)\,\mathrm{d}^3x = \mathrm{const}\,. \tag{35.6}$$

Equation (35.5) is identical to the relation (34.24) resulting from the identities established in Section 34, in which therefore we have already obtained the transcription (35.4′) of the last term of Equation (35.4). According to (34.25) the quantities $\mathfrak{t}_\lambda{}^\nu$ are quadratic-homogeneous in the first derivatives of $g_{\lambda\mu}$. They are therefore analogous to the components of the stress-energy-momentum tensor of the electromagnetic field and consequently we would expect them to represent the stresses as well as the density of energy and momentum of the gravitational field. There is, however, one important difference: The quantity $\mathfrak{t}_\lambda{}^\mu$ is not a tensor, as seen at once from the remark that all its components vanish at an arbitrary point P when we use a coordinate system which is geodesic at P. We say that $\mathfrak{t}_\lambda{}^\mu$ is a *pseudotensor* and Equation (35.5) a *pseudotensorial* equation.

In spite of its pseudotensorial character equation (35.5) can be used to determine the total energy and momentum of a system, at least in some cases and under certain conditions which we shall describe now. Equation (35.6) follows from (35.5) by an integration over e.g. a hypersurface $t=\text{const}$ if

$$\int\limits_{S\to\infty} (\mathfrak{T}^i_\mu + \mathrm{t}^i_\mu)\, n_i\, \mathrm{d}S = 0\,. \tag{35.7}$$

The assumption that we are dealing with a bounded material system ensures only that $\mathfrak{T}_\mu{}^\nu$ vanishes at some distance from the origin. The gravitational field will certainly be weak at large distances, according to the general requirement (31.2′), but it cannot vanish strictly at some finite distance. It follows that $\mathrm{t}_\mu{}^\nu$ will vanish only asymptotically for $r\to\infty$. The conservation law (35.6) will be valid if $\mathrm{t}_\mu{}^\nu$ vanishes for $r\to\infty$ so as to satisfy the condition (35.7) with $\mathfrak{T}_\mu{}^i=0$:

$$\int\limits_{S\to\infty} \mathrm{t}^i_\mu n_i\, \mathrm{d}S = 0\,. \tag{35.7′}$$

A remark of practical importance is that we have to use coordinates which are Cartesian-like, at least at large distance. Using e.g. polar coordinates we should have $\mathrm{t}_\mu{}^\nu\neq 0$ even if we were considering the Minkowski space.

We now can conclude without difficulty that the condition (35.7′) will be satisfied if the field is time-independent at large distances. Indeed, in this case with the help of the approximation method for weak fields developed in Section 31 one can derive from the field equations that the metric will be of the following form:

$$\left.\begin{aligned} &g_{\mu\nu} = \eta_{\mu\nu} + h_{\mu\nu} + O(r^{-2});\\ &h_{00} = h_{11} = h_{22} = h_{33} = \frac{-2m}{r}, \qquad h_{\mu\nu} = 0 \quad \text{for} \quad \mu \neq \nu\,. \end{aligned}\right\} \tag{35.8}$$

The Christoffel symbols will in this case be $\sim r^{-2}$. Consequently $\mathrm{t}_\lambda{}^\mu \sim r^{-4}$ and, since the area of the integration surface S is $\sim r^2$, the integrals (35.7′) vanish asymptotically when $r\to\infty$.

We shall calculate the values P_λ of this field. Because of the identity (34.31) this calculation reduces to an integration on a surface S at infinity, e.g. the surface $r= =\text{const}\to\infty$. Indeed, putting $\varrho=0$ in (34.31) we get:

$$\kappa(\mathfrak{T}^0_\lambda + \mathrm{t}^0_\lambda) = {}_{\mathrm{F}}U_\lambda{}^{0\sigma}{}_{,\sigma} = {}_{\mathrm{F}}U_\lambda{}^{0i}{}_{,i}\; (i=1,2,3)\,, \tag{35.9}$$

since ${}_{\mathrm{F}}U_\lambda{}^{00}$ vanishes because of (34.30). Therefore:

$$\kappa P_\lambda = \int\limits_{S\to\infty} {}_{\mathrm{F}}U_\lambda{}^{0i} n_i\, \mathrm{d}S\,. \tag{35.10}$$

Thus, in order to calculate P_λ we need only the asymptotic form of the metric at $r\to\infty$.

According to (34.29) ${}_{\mathrm{F}}U_\lambda{}^{0i}$ is linear and homogeneous in the first derivatives of $g_{\lambda\mu}$, which are $\sim r^{-2}$ because of (35.8). Since the surface element $\mathrm{d}S$ is $\sim r^2$, we have to replace in ${}_{\mathrm{F}}U_\lambda{}^{0i}$ all non-differentiated factors $g_{\lambda\mu}$ by $\eta_{\lambda\mu}$. The result for $\lambda=0$ is then found to be:

$$ {}_{\mathrm{F}}U_0{}^{0i} = h_{00,i} = \frac{2m}{r^2}\frac{x^i}{r}. $$

Consequently, if we put:

$$ P_0 = \int (\mathfrak{T}^0_0 + \mathfrak{t}^0_0)\, \mathrm{d}^3x \equiv Mc^2, \tag{35.11} $$

we find from (35.10):

$$ \kappa Mc^2 = 8\pi m. \tag{35.12} $$

This result is identical to (32.11). A similar calculation for ${}_{\mathrm{F}}U_l^{0i}$ ($l=1, 2, 3$) shows that the total momentum corresponding to (35.8) vanishes:

$$ P_l = 0, \tag{35.13} $$

as we should expect since we have chosen the coordinate system so that the asymptotic metric is stationary.

We only mention, without entering into the details of the calculation, that the same results follow from the form (34.38) of the conservation law.

For the total energy of a stationary gravitational field there is an interesting, entirely different formula which we shall derive now. For a stationary field we find from (34.25) and (34.24):

$$ \kappa \mathfrak{t}^0_0 = -\tfrac{1}{2}\mathfrak{L}', \quad (\mathfrak{T}^i_\lambda + \mathfrak{t}^i_\lambda)_{,i} = 0. \tag{35.14} $$

(Latin indices take the values 1, 2, 3 only.) From the last equation we derive the relation:

$$ \mathfrak{T}^i_\lambda + \mathfrak{t}^i_\lambda = \{x^i(\mathfrak{T}^k_\lambda + \mathfrak{t}^k_\lambda)\}_{,k}. \tag{35.15} $$

Integrating the last equation over a hypersurface $t=\text{const}$ we find:

$$ \int (\mathfrak{T}^i_\lambda + \mathfrak{t}^i_\lambda)\, \mathrm{d}^3x = \int_{S\to\infty} x^i(\mathfrak{T}^k_\lambda + \mathfrak{t}^k_\lambda)\, n_k\, \mathrm{d}S. $$

The last integral vanishes because for $r \to \infty$ we have:

$$ \mathfrak{T}^k_\lambda = 0, \qquad \mathfrak{t}^k_\lambda \sim \frac{1}{r^4}. $$

Therefore we have, in the case of a stationary field:

$$ \int (\mathfrak{T}^i_\lambda + \mathfrak{t}^i_\lambda)\, \mathrm{d}^3x = 0. \tag{35.16} $$

Since $\mathfrak{L}'$ is quadratic-homogeneous in the first derivatives of $\mathfrak{g}^{\alpha\beta}$, we shall have:

$$ \mathfrak{g}^{\alpha\beta}{}_{,\lambda}\mathfrak{L}'^{\lambda}_{\alpha\beta} = 2\mathfrak{L}'. \tag{35.17} $$

Therefore from (34.25):

$$\kappa \mathfrak{t}_\lambda{}^\lambda = -\mathfrak{L}' . \tag{35.18}$$

Combining (35.18) with the first (35.14) we find for a stationary field:

$$\mathfrak{t}_0^0 - \mathfrak{t}_i^i = 0 . \tag{35.19}$$

We now write, according to (35.6) and (35.16):

$$P_0 = \int (\mathfrak{T}_0^0 + \mathfrak{t}_0^0)\, \mathrm{d}^3x - \int (\mathfrak{T}_i^i + \mathfrak{t}_i^i)\, \mathrm{d}^3x .$$

Therefore, because of (35.19):

$$P_0 = \int (\mathfrak{T}_0^0 - \mathfrak{T}_i^i)\, \mathrm{d}^3x . \tag{35.20}$$

The interesting feature of this relation is that it expresses the total energy of a stationary system as an integral over the material sources of the field.

An equation expressing the conservation law of angular momentum can be derived from the identities (34.24) and (34.31). It is, however, simpler to formulate this conservation law using the Landau-Lifshitz identities, as we shall show now. Indeed, if we introduce the abreviation:

$$\Sigma^{\mu\nu} = \sqrt{-g}\,(\mathfrak{T}^{\mu\nu} + \tilde{\mathfrak{t}}^{\mu\nu}), \tag{35.21}$$

we arrive at the conservation law of angular momentum in the same way as in special relativity, the only difference being that we shall now use $\Sigma^{\mu\nu}$ instead of the special relativistic $T^{\mu\nu}$.

We define a quantity $\mathfrak{M}^{\lambda\mu\nu}$ by the relation:

$$\mathfrak{M}^{\lambda\mu\nu} = x^\lambda \Sigma^{\mu\nu} - x^\mu \Sigma^{\lambda\nu} . \tag{35.22}$$

We then verify directly the equation:

$$\mathfrak{M}^{\lambda\mu\nu}{}_{,\nu} = 0 \tag{35.23}$$

which is a consequence of (34.38) and of the symmetry of $\Sigma^{\mu\nu}$ in μ, ν.

Equation (35.23) is the differential form of the conservation law of angular momentum. The integral form is found by integrating (35.23) over a spacelike hypersurface, e.g. a hypersurface $t=\text{const}$:

$$\frac{\mathrm{d}}{\mathrm{d}x^0} \int_{t=\text{const}} \mathfrak{M}^{\lambda\mu 0}\, \mathrm{d}^3x + \int_{S\to\infty} \mathfrak{M}^{\lambda\mu i} n_i\, \mathrm{d}S = 0 . \tag{35.24}$$

For a bounded material system we have in special relativity (with $\Sigma^{\mu\nu}$ replaced by $T^{\mu\nu}$):

$$\int_{S\to\infty} \mathfrak{M}^{\lambda\mu i} n_i\, \mathrm{d}S = 0 . \tag{35.25}$$

We shall have the same result in general relativity if $\tilde{\mathfrak{t}}^{\mu\nu}$ vanishes sufficiently fast as $r \to \infty$. This is the case if the field is time-independent at large distances. Equation (35.24) reduces then to the integral conservation law

$$cJ^{\lambda\mu} \equiv \int\limits_{t=\text{const}} \mathfrak{M}^{\lambda\mu 0} \, d^3x = \int\limits_{t=\text{const}} (x^\lambda \Sigma^{\mu 0} - x^\mu \Sigma^{\lambda 0}) \, d^3x = \text{const}. \tag{35.26}$$

The components of the angular momentum are the quantities J^{ik}:

$$\mathbf{J} = (J^{23}, J^{31}, J^{12}).$$

The components J^{0i} express the law of uniform motion of a generalized 'centre of mass' (or better, centre of energy) of the system.

With the help of Equation (34.32) and (34.35) we can easily write $\mathfrak{M}^{\lambda\mu\nu}$ in the form:

$$\mathfrak{M}^{\lambda\mu\nu} = \mathfrak{F}^{\lambda\mu\nu\varrho}{}_{,\varrho}\,; \qquad \mathfrak{F}^{\lambda\mu\nu\varrho} = -\,\mathfrak{F}^{\lambda\mu\varrho\nu}. \tag{35.27}$$

We then rewrite Equation (35.26) as follows:

$$cJ^{\lambda\mu} = \int\limits_{S\to\infty} \mathfrak{F}^{\lambda\mu 0 i} n_i \, dS. \tag{35.28}$$

This relation shows that the values of $J^{\lambda\mu}$ depend only on the asymptotic form of the metric.

The discussion of the integrals appearing in (35.28) shows that the part (35.8) of the metric does not contribute to the quantities $J^{\lambda\mu}$. Actually the components J^{ik} are found to depend only on the terms of the metric which are $\sim r^{-2}$ and have the form (32.29). The detailed calculation confirms the result (32.28).

Exercises

VIII1: Show that, if one uses instead of (33.29) the Lagrangian

$$\mathfrak{L} = \mathfrak{g}^{\lambda\mu}(R_{\lambda\mu} + \varkappa T_{\lambda\mu})$$

and keeps the term $T_{\lambda\mu}$ unvaried, one arrives at the field equation (19.4′).

VIII2: Determine the expression giving the quantity $\mathfrak{F}^{\lambda\mu\nu\varrho}$ in Equation (35.27).

VIII3: Using Equation (35.28) and the expression for $\mathfrak{F}^{\lambda\mu\nu\varrho}$ (see preceding exercise) show that for a stationary metric containing terms of order $1/r^2$ of the form (32.29) the coefficients J^k are the components of the total angular momentum.

CHAPTER IX

THE EINSTEIN-MAXWELL EQUATIONS

36. Gravitational and Electromagnetic Field

In special relativity, when we use inertial coordinates, we have the Maxwell equations in the form:

$$\eta^{\mu\nu} F_{\lambda\mu,\nu} = s_\lambda\,; \qquad F_{[\lambda\mu,\nu]} = 0\,, \tag{36.1}$$

$\eta^{\mu\nu}$ being the Minkowski metric (31.1). The second equation allows the introduction of the vector potential A_λ. Indeed, this equation is satisfied identically if we put:

$$F_{\lambda\mu} = A_{\mu,\lambda} - A_{\lambda,\mu}\,. \tag{36.2}$$

If we allow arbitrary coordinates in Minkowski space, we have to write the first Equation (36.1) in its covariant form:

$$F^{\lambda\mu}{}_{;\mu} = s^\lambda\,; \qquad F^{\lambda\mu} = \gamma^{\lambda\varrho}\gamma^{\mu\sigma} F_{\varrho\sigma}\,, \qquad s^\lambda = \gamma^{\lambda\varrho} s_\varrho\,, \tag{36.3}$$

where $\gamma^{\lambda\mu}$ is the Minkowski metric in these coordinates. According to (4.13′) we can write Equation (36.3) also in the form:

$$(\sqrt{-\gamma}\, F^{\lambda\mu})_{,\mu} = \sqrt{-\gamma}\, s^\lambda\,, \tag{36.3′}$$

where γ is the determinant of $\gamma_{\lambda\mu}$. The second Equation (36.1) is already covariant, according to (4.8).

In order to rewrite the first Maxwell equation for a Riemannian space we have to replace in (36.3) or (36.3′) the Minkowski metric $\gamma^{\lambda\mu}$ by the metric $g^{\lambda\mu}$ of the Riemannian space:

$$(\sqrt{-g}\, F^{\lambda\mu})_{,\mu} = \sqrt{-g}\, s^\lambda\,; \qquad F^{\lambda\mu} = g^{\lambda\varrho} g^{\mu\sigma} F_{\varrho\sigma}\,, \qquad s^\lambda = g^{\lambda\varrho} s_\varrho\,. \tag{36.4}$$

It follows that in general relativity, i.e. in the presence of a gravitational field, the equations of the electromagnetic field will be given by (36.4) and (36.2). Thus the use of the metric $g_{\lambda\mu}$ in (36.4) expresses the influence of the gravitational field on the electromagnetic one.

Is there an influence of the electromagnetic on the gravitational field and how shall we describe it? This question also can be answered in an elementary manner with the help of known results from special relativity. Let us consider a charged material system being described by the phenomenological matter tensor $T_{\lambda\mu}$ and the charge distribution s^λ. The electromagnetic interaction is taken into account correctly if we write the conservation law of energy and momentum in the form:

$$\eta^{\mu\nu}(T_{\lambda\mu} + {}_M T_{\lambda\mu})_{,\nu} = 0\,. \tag{36.5}$$

The tensor ${}_{\mathrm{M}}T_{\lambda\mu}$ is the stress-energy-momentum tensor of the Maxwell field:

$$ {}_{\mathrm{M}}T_{\lambda}^{\mu} = - F_{\lambda\alpha}F^{\mu\alpha} + \tfrac{1}{4}\delta_{\lambda}^{\mu}F_{\alpha\beta}F^{\alpha\beta} . \tag{36.6} $$

Indeed, if we calculate the expression:

$$ \eta^{\mu\nu}{}_{\mathrm{M}}T_{\lambda\mu,\nu} \equiv {}_{\mathrm{M}}T_{\lambda}{}^{\nu}{}_{,\nu} , $$

we find with the help of the Maxwell equations (36.1):

$$ {}_{\mathrm{M}}T_{\lambda}{}^{\nu}{}_{,\nu} = - F_{\lambda\nu}s^{\nu} . \tag{36.7} $$

We therefore have from (36.5):

$$ T_{\lambda}{}^{\nu}{}_{,\nu} = F_{\lambda\nu}s^{\nu} . \tag{36.8} $$

The right-hand side of this equation is the Lorentz force-density and consequently Equations (36.5) are the correct equations of motion of the system.

It follows from this remark that in general relativity we have to replace $T_{\lambda\mu}$ by $T_{\lambda\mu} + {}_{\mathrm{M}}T_{\lambda\mu}$. So the field equations of the gravitational field will be:

$$ R_{\lambda\mu} - \tfrac{1}{2}g_{\lambda\mu}R = - \kappa(T_{\lambda\mu} + {}_{\mathrm{M}}T_{\lambda\mu}) . \tag{36.9} $$

The influence of the electromagnetic on the gravitational field is expressed by the presence of the term ${}_{\mathrm{M}}T_{\lambda\mu}$ in the sources of the gravitational field. We thus have derived in an elementary way the field equations of the combined gravitational and electromagnetic field, which are the Equations (36.9) and (36.4).

The field equations (36.9) and (36.4) are seen immediately to lead to the following conservation laws:

$$ (T_{\lambda}^{\mu} + {}_{\mathrm{M}}T_{\lambda}^{\mu})_{;\mu} = 0 , \tag{36.10} $$

$$ s^{\lambda}{}_{;\lambda} = 0 . \tag{36.11} $$

It is easy to prove that Equation (36.7) remains valid also in the Riemannian space, but in its covariant form:

$$ {}_{\mathrm{M}}T_{\lambda}{}^{\mu}{}_{;\mu} = - F_{\lambda\mu}s^{\mu} . \tag{36.12} $$

Therefore Equation (36.10) is equivalent to:

$$ T_{\lambda}{}^{\mu}{}_{;\mu} = F_{\lambda\mu}s^{\mu} . \tag{36.13} $$

The pseudotensorial conservation laws derived in Section 35 will now contain the sum $T_{\lambda}^{\mu} + {}_{\mathrm{M}}T_{\lambda}^{\mu}$ instead of T_{λ}^{μ}.

The field equations (36.9) and (36.4) can be derived from a variational principle. We shall consider here only the case of the 'electrovacuum', characterized by the conditions:

$$ T_{\lambda\mu} = 0 , \qquad s_{\lambda} = 0 . \tag{36.14} $$

The appropriate Lagrangian is then simply the sum of the Lagrangians of the gravitational and the electromagnetic field, the last one written for a Riemannian space with the metric $g_{\lambda\mu}$:

$$\mathfrak{L} = \mathfrak{L}_{gr} + \alpha \mathfrak{L}_{em} ; \tag{36.15}$$

$$\mathfrak{L}_{gr} \equiv \mathfrak{L}' = \mathfrak{g}^{\lambda\mu} (\Gamma^{\alpha}_{\lambda\mu} \Gamma^{\beta}_{\alpha\beta} - \Gamma^{\alpha}_{\lambda\beta} \Gamma^{\beta}_{\mu\alpha}), \tag{36.16}$$

$$\mathfrak{L}_{em} = \frac{1}{\sqrt{-g}} \mathfrak{g}^{\lambda\varrho} \mathfrak{g}^{\mu\sigma} F_{\lambda\mu} F_{\varrho\sigma} . \tag{36.17}$$

The constant factor α in (36.15) is needed because the Lagrangians (36.16) and (36.17) have different physical dimensions:

$$\mathfrak{L}_{gr} \sim (\text{length})^{-2}, \qquad \mathfrak{L}_{em} \sim \text{energy} \times (\text{length})^{-3}.$$

The value of α is determined from the requirement that the equations of the gravitational field reduce to (36.9) simplified by the first of conditions (36.14). This value is:

$$\alpha = -\frac{\kappa}{2}, \tag{36.18}$$

where κ is the relativistic gravitational constant appearing in Equation (36.9).

The Lagrangian (36.15) contains the field functions $\mathfrak{g}^{\lambda\mu}$ and A_λ. Since $\mathfrak{L}$ contains only first derivatives of the field functions, the resulting field equations will be of the form (33.10). The variation of the gravitational field variables $\mathfrak{g}^{\lambda\mu}$ leads to the gravitational field equations:

$$\frac{\partial \mathfrak{L}}{\partial \mathfrak{g}^{\lambda\mu}} - \frac{\partial}{\partial x^\nu} \left(\frac{\partial \mathfrak{L}}{\partial \mathfrak{g}^{\lambda\mu}{}_{,\nu}} \right) = 0 . \tag{36.19}$$

Similarly the variation of the electromagnetic variables A_λ will lead to the electromagnetic field equations:

$$\frac{\partial \mathfrak{L}}{\partial A_\lambda} - \frac{\partial}{\partial x^\nu} \left(\frac{\partial \mathfrak{L}}{\partial A_{\lambda,\nu}} \right) = 0 . \tag{36.20}$$

The discussion of the gravitational Lagrangian $\mathfrak{L}_{gr}$ has led to the relation (33.23):

$$\frac{\partial \mathfrak{L}_{gr}}{\partial \mathfrak{g}^{\lambda\mu}} - \frac{\partial}{\partial x^\nu} \left(\frac{\partial \mathfrak{L}_{gr}}{\partial \mathfrak{g}^{\lambda\mu}{}_{,\nu}} \right) = R_{\lambda\mu} . \tag{36.21}$$

Since $\mathfrak{L}_{gr}$ does not contain A_λ we have:

$$\frac{\partial \mathfrak{L}_{gr}}{\partial A_\lambda} = 0 = \frac{\partial \mathfrak{L}_{gr}}{\partial A_{\lambda,\mu}} . \tag{36.22}$$

There remains to calculate the expression for $\delta \mathfrak{L}_{em}$ and to determine the derivatives of $\mathfrak{L}_{em}$ with respect to the field variables and their derivatives. The calculation is elementary and leads to the following results:

$$\left.\begin{aligned}
&\frac{\partial \mathfrak{L}_{\text{em}}}{\partial \mathfrak{g}^{\lambda\mu}} = 2\left(g^{\alpha\beta} F_{\lambda\alpha} F_{\mu\beta} - \tfrac{1}{4} g_{\lambda\mu} F^{\alpha\beta} F_{\alpha\beta}\right) = -\, 2\,{}_{\mathrm{M}}T_{\lambda\mu}\,,\\
&\frac{\partial \mathfrak{L}_{\text{em}}}{\partial \mathfrak{g}^{\lambda\mu}{}_{,\nu}} = 0;\\
&\frac{\partial \mathfrak{L}_{\text{em}}}{\partial A_{\mu}} = 0\,, \qquad \frac{\partial \mathfrak{L}_{\text{em}}}{\partial A_{\mu,\nu}} = 4\sqrt{-g}\; F^{\nu\mu}\,.
\end{aligned}\right\} \tag{36.23}$$

The field equations (36.19) and (36.20) are then found to be:

$$R_{\lambda\mu} + \kappa\,{}_{\mathrm{M}}T_{\lambda\mu} = 0\,, \tag{36.24}$$

$$\left(\sqrt{-g}\; F^{\lambda\mu}\right)_{,\mu} = 0\,. \tag{36.25}$$

Equation (36.25) is identical with (36.4), simplified by the second of the conditions (36.14). Equation (36.24) is actually equivalent to (36.9) simplified by the first of Equations (36.14). In order to prove this, one has to take into account the relation resulting from the contraction of (36.9) by $g^{\lambda\mu}$. The trace of ${}_{\mathrm{M}}T_{\lambda\mu}$ vanishes:

$$g^{\lambda\mu}\;{}_{\mathrm{M}}T_{\lambda\mu} = 0\,, \tag{36.26}$$

and consequently we find in general:

$$R = \kappa T\,, \qquad T \equiv g^{\lambda\mu} T_{\lambda\mu}\,. \tag{36.27}$$

Therefore we have now, because of the first condition (36.14):

$$R = 0 \tag{36.28}$$

and consequently Equation (36.9) reduces in this case to (36.24).

37. The Reissner-Nordström Solution

As an application of the field equations (36.24) and (36.25) we shall determine their spherically symmetric solution, the *Reissner-Nordström solution*. We shall consider time-dependent functions $g_{\lambda\mu}$ and A_{λ}, in order to establish the generalized Birkhoff theorem for the combined gravitational and electromagnetic field.

The spherically symmetric $g_{\lambda\mu}$ was discussed in Section 20, where we showed that it can be reduced to the form (20.15):

$$\mathrm{d}s^2 = e^{\nu}\,\mathrm{d}t^2 - e^{\mu}\,\mathrm{d}r^2 - r^2\left(\mathrm{d}\theta^2 + \sin^2\theta\,\mathrm{d}\varphi^2\right), \tag{37.1}$$

ν and μ being functions of t and r. A spherically symmetric vector A_{λ} is seen at once to have, in the polar coordinates used in (37.1), vanishing components A_2 and A_3; the components A_0 and A_1 can depend on t and r only:

$$A_0 = \varphi(t, r)\,, \qquad A_1 = f(t, r)\,; \qquad A_2 = A_3 = 0\,. \tag{37.2}$$

The electromagnetic vector potential A_λ is determined up to a gauge transformation:

$$\tilde{A}_\lambda = A_\lambda + \Lambda_{,\lambda}, \tag{37.3}$$

Λ being an arbitrary function. Using a function Λ which satisfies the relation

$$\Lambda_{,1} \equiv \frac{\partial \Lambda}{\partial r} = - f(t, r)$$

we can make the component $\tilde{A}_1$ vanish. We therefore assume A_λ to be of the simpler form:

$$A_0 = \varphi(t, r); \qquad A_i = 0. \tag{37.4}$$

The calculation of the electromagnetic field $F_{\lambda\mu}$ from (36.2) and (37.4) leads to the following simple results:

$$F_{01} = - F_{10} = - \varphi', \quad \text{all other} \quad F_{\lambda\mu} = 0; \qquad \varphi' \equiv \frac{\partial \varphi}{\partial r}. \tag{37.5}$$

Using the metric (37.1) we find from the second of Equations (36.4):

$$F^{01} = - F^{10} = e^{-(\mu+\nu)} \varphi', \quad \text{all other} \quad F^{\lambda\mu} = 0. \tag{37.6}$$

Again from the metric (37.1) we have:

$$\sqrt{-g} = r^2 e^{(\mu+\nu)/2} \sin\theta. \tag{37.7}$$

The Maxwell equation (36.25) gives now the following two relations:

$$(\sqrt{-g}\, F^{01})' = 0 = (\sqrt{-g}\, F^{01})^{\cdot}; \qquad (\,.\,.)^{\cdot} \equiv \frac{\partial}{\partial t}(\,.\,.). \tag{37.8}$$

These equations are integrated at once, giving the result:

$$e^{-(\mu+\nu)/2}\, r^2 \varphi' = - e, \qquad e = \text{const.} \tag{37.9}$$

The comparison with the solution in the Minkowski space shows that the integration constant e is the total charge of the interior material system which is the source of this field.

We now consider the gravitational field equations (36.24). We start with the component $\lambda = 0$, $\mu = 1$ of this equation. The quantity R_{01} is given by Equation (21.17):

$$R_{01} = - \frac{1}{r} \dot{\mu}.$$

From (37.5) and the first of (36.23) we find:

$${}_{\mathrm{M}}T_{01} = 0.$$

It follows that the component $\lambda = 0$, $\mu = 1$ of Equation (36.24) reduces to:

$$\dot{\mu} = 0, \tag{37.10}$$

which is identical with (21.18). We then conclude, in the same way as in Section 21, that the non-vanishing $R_{\lambda\mu}$ are given by the formulae (21.6).

In order to obtain the corresponding components of the field equations (36.24) we have to calculate the components ${}_{\mathrm{M}}T_{00}$, ${}_{\mathrm{M}}T_{11}$ and ${}_{\mathrm{M}}T_{22}$ from (37.5), (37.6) and the first Equation (36.23). A simple calculation leads finally to the following equations:

$$e^{\mu-\nu}R_{00} = -\frac{\nu''}{2} - \frac{\nu'}{r} + \nu'\frac{\mu'-\nu'}{4} = -\frac{\kappa}{2}e^{-\nu}\varphi'^2, \tag{37.11}$$

$$R_{11} = \frac{\nu''}{2} - \frac{\mu'}{r} - \nu'\frac{\mu'-\nu'}{4} = \frac{\kappa}{2}e^{-\nu}\varphi'^2, \tag{37.12}$$

$$R_{22} = e^{-\mu}\left(1 - e^{\mu} - r\frac{\mu'-\nu'}{2}\right) = -\frac{\kappa}{2}e^{-(\mu+\nu)}r^2\varphi'^2. \tag{37.13}$$

The sum of Equations (37.11) and (37.12) is identical with the simple Equation (21.9):

$$\mu' + \nu' = 0. \tag{37.14}$$

Introducing (37.14) and (37.9) into (37.13) we find:

$$e^{-\mu} - 1 - r\mu' e^{-\mu} = -\frac{\kappa\, e^2}{2\, r^2}.$$

We can rewrite this last equation in the following form:

$$\left(re^{-\mu} - r - \frac{\kappa e^2}{2r}\right)' = 0. \tag{37.15}$$

Integrating we find:

$$re^{-\mu} - r - \frac{\kappa e^2}{2r} = -2m, \tag{37.16}$$

where m is independent of r. Since the left-hand side of (37.16) is, according to (37.10), independent of t, m is a constant. We thus have:

$$e^{-\mu} = 1 - \frac{2m}{r} + \frac{\kappa e^2}{2r^2}. \tag{37.17}$$

Finally we determine the function $\nu(t, r)$ from (37.14):

$$\nu = -\mu + \lambda \quad \Leftrightarrow \quad e^{\nu} = e^{-\mu+\lambda}, \tag{37.18}$$

λ being an arbitrary function of t. This completes the integration of the field equations. Indeed, one can verify directly that both Equations (37.11) and (37.12) are now satisfied.

Because of the factor e^{λ} in e^{ν} the metric is time-dependent, but only apparently:

As we have remarked in Section 21, the transformation $t \to t'$ of the time coordinate, which is determined by the relation

$$e^{\lambda/2}\,\mathrm{d}t = \mathrm{d}t',$$

reduces λ to zero. Writing again t instead of t' we have the following final form of the solution:

$$e^{\nu} = e^{-\mu} = 1 - \frac{2m}{r} + \frac{\kappa e^2}{2r^2}; \tag{37.19}$$

$$F_{01} = -F_{10} = \frac{e}{r^2} = -F^{01} = +F^{10}; \quad A_0 = \frac{e}{r}. \tag{37.20}$$

The solution contains the two integration constants m and e. The constant m is the total energy of the system in geometrical units. This follows from the general result (35.11) derived from the conservation law of energy and momentum. The second constant e is the total charge of the system, as is shown clearly by the last formula (37.20). The fact that we arrived at a time-independent solution, though we allowed at the beginning time-dependent field variables, constitutes the proof of the generalized Birkhoff theorem for the combined gravitational and electromagnetic field.

The energy density of the Maxwell field is found from (36.6) and (37.20):

$${}_{\mathrm{M}}T^0_0 = -F_{0\alpha}F^{0\alpha} + \tfrac{1}{4}F_{\alpha\beta}F^{\alpha\beta} = -\tfrac{1}{2}F_{01}F^{01} = \frac{e^2}{2r^4}. \tag{37.21}$$

It is given by exactly the same formula as in the Minkowski space (special relativity).

The *exterior* solution determined by (37.19) and (37.20) will be valid in the region $r > a(t)$, if $r = a(t)$ is the boundary of the material distribution which is the source of the field. Determining the *interior* solution in the domain $r < a(t)$ is again an essentially more complicated problem which we shall not discuss here.

38. The Singularity and the Horizon of the Reissner-Nordström Solution

In this section we shall discuss in some detail the case in which the Reissner-Nordström solution is assumed valid for all positive values of r. The interest in this case lies in the fact that, when the constants m and e satisfy a certain condition, this solution also has a horizon.

Figure 22 represents the quantity g_{00} as a function of r. For comparison we show on the figure also the corresponding quantity ${}_{s}g_{00}$ of the Schwarzschild solution:

$${}_{s}g_{00} = 1 - \frac{2m}{r}.$$

The function $g_{00}(r)$ will vanish for the values of r which are the roots of the equation:

$$r^2 - 2mr + \frac{\kappa e^2}{2} = 0. \tag{38.1}$$

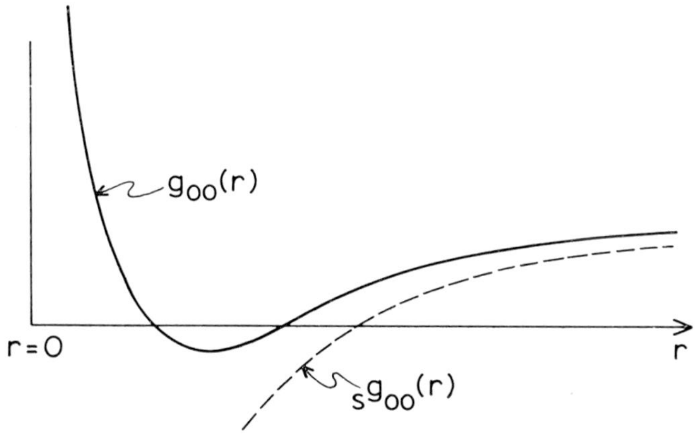

Fig. 22. The quantity g_{00} as a function of r.

Equation (38.1) will have two distinct roots $r_1 > r_2$ if

$$m > |e| \sqrt{\frac{\kappa}{2}}. \tag{38.2}$$

This is the case shown on Figure 22. There will be one double root or no real root if

$$m = |e| \sqrt{\frac{\kappa}{2}} \quad \text{or} \quad m < |e| \sqrt{\frac{\kappa}{2}}.$$

We shall limit ourselves to the discussion of the case (38.2).

The quantity g_{00} can be written in the form:

$$g_{00} = \frac{1}{r^2} (r - r_1)(r - r_2). \tag{38.3}$$

For $r = r_1 = (OA)$ we have a situation similar to the one we had in the Schwarzschild case for $r = 2m$: g_{00} vanishes at A and has different sign on either side of A. It follows that the parametric lines of the t-coordinate (i.e. the lines on which r, θ and φ are constant) are time-like for $r > r_1$, null for $r = r_1$, and space-like for $r < r_1$. The difference with the Schwarzschild metric is that we now have the second point B at the distance $r = r_2$: The parametric lines of t are again time-like for $r < r_2$.

One can verify at once that, as we found in Section 26 for the Schwarzschild case, the parametric lines of the coordinate r are geodesics. Indeed, this follows immediately from the remarks that $g_{\lambda\mu}$ is diagonal and g_{11} depends on r only. These geodesics have a different character – space-like or time-like – on either side of each one of the points A and B and consequently they cannot form one continuous geodesic. The orientation of the light cones is also different on the two sides of the points A and B, in a way similar to that shown on Figure 8. It follows from these remarks that the coordinates used in the metric (37.1) are not meaningful at the points A and B.

We can perform again, as we did in the Schwarzschild case, a transformation leading to Eddington-type coordinates. For $r > r_1$ the transformation will be of the following form:

$$t = \tilde{t} + \alpha_1 \lg(r - r_1) + \alpha_2 \lg(r - r_2), \quad \alpha_1 \text{ and } \alpha_2 \text{ constants}, \tag{38.4}$$

the coordinates r, θ and φ remaining unchanged. We then have:

$$\mathrm{d}t = \mathrm{d}\tilde{t} + \left(\frac{\alpha_1}{r - r_1} + \frac{\alpha_2}{r - r_2} \right) \mathrm{d}r . \tag{38.5}$$

Introducing this expression into the metric (37.1) and (37.19) we find:

$$\left.\begin{aligned} \mathrm{d}s^2 = {} & \frac{(r - r_1)(r - r_2)}{r^2} \, \mathrm{d}\tilde{t}^2 + \\ & + 2 \frac{\alpha_1 (r - r_2) + \alpha_2 (r - r_1)}{r^2} \, \mathrm{d}\tilde{t} \, \mathrm{d}r + \\ & + \frac{1}{(r - r_1)(r - r_2)} \times \\ & \times \left\{ \frac{[\alpha_1 (r - r_2) + \alpha_2 (r - r_1)]^2}{r^2} - r^2 \right\} \mathrm{d}r^2 - \\ & - r^2 (\mathrm{d}\theta^2 + \sin^2\theta \, \mathrm{d}\varphi^2) . \end{aligned}\right\} \tag{38.6}$$

The following choice of the constants α_1 and α_2 eliminates the denominator $(r - r_1)(r - r_2)$ of the coefficient of $\mathrm{d}r^2$ in (38.6):

$$\alpha_1 = \frac{-r_1^2}{r_1 - r_2}, \qquad \alpha_2 = \frac{r_2^2}{r_1 - r_2} . \tag{38.7}$$

We then have:

$$\alpha_1 (r - r_2) + \alpha_2 (r - r_1) = - r (r_1 + r_2) + r_1 r_2$$

and we find finally:

$$\mathrm{d}s^2 = (1 - f) \, \mathrm{d}\tilde{t}^2 - 2f \, \mathrm{d}\tilde{t} \, \mathrm{d}r - (1 + f) \, \mathrm{d}r^2 - r^2 (\mathrm{d}\theta^2 + \sin^2\theta \, \mathrm{d}\varphi^2) ; \tag{38.8}$$

$$f = 1 - g_{00} = \frac{2m}{r} - \frac{\kappa e^2}{2r^2} . \tag{38.9}$$

One can verify immediately that, if we take instead of (38.4):

$$t = \tilde{t} + \alpha_1 \lg(r_1 - r) + \alpha_2 \lg(r - r_2) \quad \text{for} \quad r_2 < r < r_1 ,$$
$$t = \tilde{t} + \alpha_1 \lg(r_1 - r) + \alpha_2 \lg(r_2 - r) \quad \text{for} \quad r < r_2 ,$$

the relation (38.5) remains valid and consequently we obtain again the metric (38.8). One verifies immediately that the metric (38.8) is regular for all positive values of r. At the point $r = 0$ we have a true singularity, as it can be shown by the calculation of the scalar (27.1).

The general form of the radial null geodesics corresponding to the form (38.8) of the Reissner-Nordström metric will be determined if we put:

$$d\theta = d\varphi = 0 \quad \text{and} \quad ds^2 = 0.$$

The result is:

$$d\tilde{t}^2 - dr^2 - f(d\tilde{t} + dr)^2 = (d\tilde{t} + dr)\{(1-f)\,d\tilde{t} - (1+f)\,dr\} = 0.$$

Thus the one family of null geodesics has the equation:

$$\tilde{t} + r = \text{const}. \tag{38.10}$$

The second family has the differential equation:

$$\frac{d\tilde{t}}{dr} = \frac{1+f}{1-f}. \tag{38.11}$$

The diagrams on Figure 23 show the quantities $1-f$ and $1+f$. In Figure 24 we show the general form of the two types of radial null geodesics corresponding to the Equations (38.10) and (38.11). We also show the resulting orientation of the light cones at different values of r.

From the orientation of the light cone at $r = r_1$ we conclude at once that no physical particle or light signal can cross outwards the surface $r = r_1$. This surface constitutes therefore a *horizon*, similar to the one we had in the Schwarzschild case. Moreover, we see on Figure 24 that in the interval $r_2 < r < r_1$ the light cones are inclined towards the worldline of the centre of symmetry $r = 0$. It follows that any particle crossing the surface $r = r_1$ inwards will move necessarily towards the centre of symmetry until either it crosses or it reaches asymptotically the surface $r = r_2$. Note that the surface $r = r_2$ is also a characteristic hypersurface, as is the horizon: We have a situation which is qualitatively similar to the one we found in the case of the Kerr metric (Section 30).

In the region $r < r_2$ the light cones are no longer inclined towards the centre and consequently physical particles do not need to fall on the singularity. Actually neutral test particles cannot reach the singularity, as we shall show in detail now.

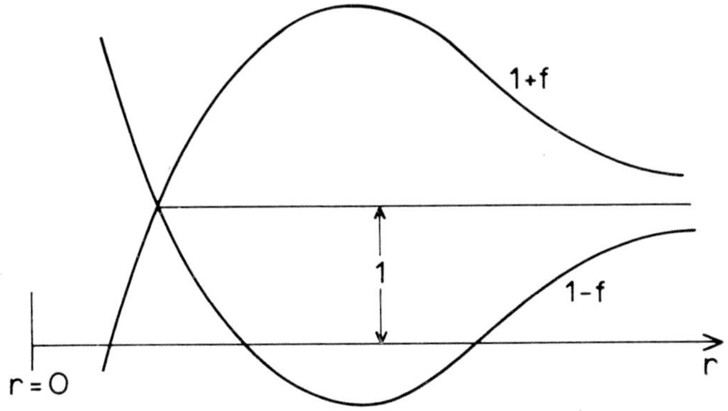

Fig. 23. The quantities $1-f$ and $1+f$ as functions of r.

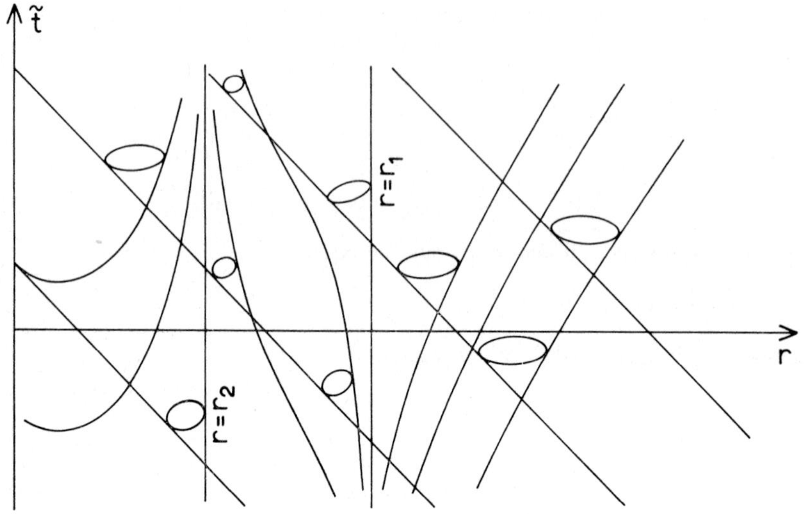

Fig. 24. Radial null geodesics and light cones of the metric (38.8).

The electromagnetic field does not act directly on a neutral test particle, which therefore is moving on a geodesic also in the present case. We limit the discussion to radial time-like geodesics. The result, which we shall obtain, can be shown easily to be valid also for non-radial time-like geodesics. Radial null-geodesics do fall on the singularity, as is shown in Figure 24.

A radial geodesic is characterized by

$$\mathrm{d}\theta = \mathrm{d}\varphi = 0 .$$

In the coordinates $\tilde{t}$, r used in the metric (38.8) we find from (13.3):

$$1 = \{g_{00} + 2g_{01}\dot{r} + g_{11}\dot{r}^2\} (u^0)^2 ; \tag{38.12}$$

$$u^0 \equiv \frac{\mathrm{d}\tilde{t}}{\mathrm{d}s}; \qquad \dot{r} \equiv \frac{\mathrm{d}r}{\mathrm{d}\tilde{t}}; \qquad u^1 = \frac{\mathrm{d}r}{\mathrm{d}s} = \dot{r} u^0 . \tag{38.13}$$

We now recall the general result (22.9) according to which in a stationary field we have the following first integral of the geodesic equation:

$$u_0 = \text{const} . \tag{38.14}$$

The relation between u_0 and u^0 is:

$$u_0 = g_{0\alpha} u^\alpha = (g_{00} + g_{01}\dot{r})\, u^0 . \tag{38.15}$$

Eliminating u^0 between (38.12) and (38.15) we obtain an equation for $\dot{r}$ which can be written finally in the form:

$$\left\{\dot{r} + \frac{fY}{1 + (1+f)\,Y}\right\}^2 = \frac{Y u_0^2}{\{1 + (1+f)\,Y\}^2}; \qquad Y \equiv u_0^2 - g_{00} . \tag{38.16}$$

On the diagram of Figure 22 let us draw a line parallel to the r-axis at a distance u_0^2 from it (Figure 25). If $u_0^2<1$, this line determines on the curve representing $g_{00}(r)$ two points A', B' situated respectively in the regions $r>r_1$ and $r<r_2$. Equation (38.16) shows that we shall have

$$\dot{r}=0 \quad \text{for} \quad r=r_{A'} \quad \text{or} \quad r=r_{B'}. \tag{38.17}$$

Moreover, values of r larger than $r_{A'}$ or smaller than $r_{B'}$ cannot be reached, since this would make the right-hand side of (38.16) negative. It follows that the motion of the test particle is, speaking intuitively, an oscillation between the extreme points A' and B'. When we have $u_0^2>1$ there is only the minimum distance $r_{B''}$ (Figure 25) and when $u_0^2\to\infty$ the minimum distance $r_{B''}\to 0$: This is the case of the radial null geodesic which reaches the singularity, as we mentioned earlier.

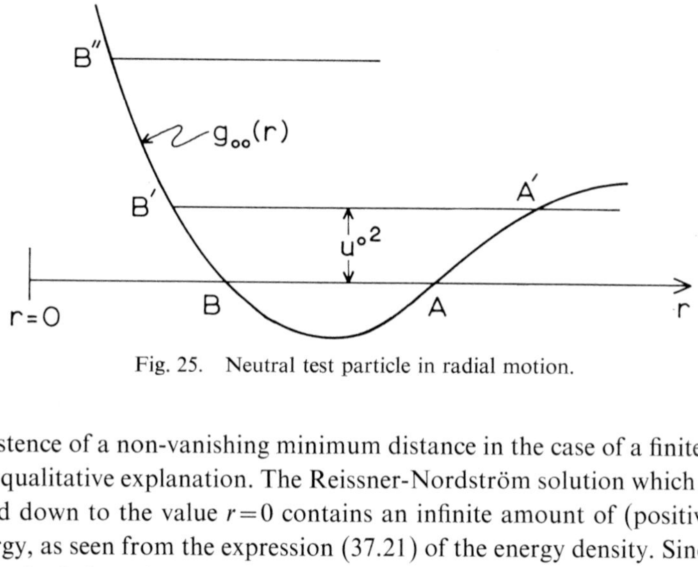

Fig. 25. Neutral test particle in radial motion.

The existence of a non-vanishing minimum distance in the case of a finite u_0 has the following qualitative explanation. The Reissner-Nordström solution which is assumed to be valid down to the value $r=0$ contains an infinite amount of (positive) electrostatic energy, as seen from the expression (37.21) of the energy density. Since the total energy has the finite value m, there will be necessarily an infinite amount of negative (non-electrostatic) energy at the centre of symmetry. It is because of this negative energy that at sufficiently small distances r the gravitational force is no longer attractive, but repulsive. This remark shows that this solution is, at least in its classical form, highly unphysical.

The more detailed discussion of the motion of the particle in the interval $r_{B'}<r<r_{A'}$ leads to another interesting conclusion. Solving Equation (38.16) for $\dot{r}$ we find:

$$\dot{r}=\frac{-fY\pm u_0\sqrt{Y}}{1+(1+f)Y}. \tag{38.18}$$

The double sign corresponds to the two possibilities $\dot{r}>0$ or $\dot{r}<0$. The sign $-$ corresponds to the particle moving inwards, from A' to B', and the sign $+$ to the particle moving outwards after having reached the point B'. With the sign $-$ we find that $\dot{r}$

is negative in the whole interval $r_{B'} < r < r_{A'}$. In particular at the points $r = r_1$ or $r = r_2$, where $g_{00} = 0$ or $f = 1$, we find the value

$$\dot{r}_- = \frac{-2u_0^2}{1 + 2u_0^2}. \tag{38.19}$$

On the contrary, with the sign $+$ we find at the points $r = r_1$ or r_2:

$$\dot{r}_+ = 0. \tag{38.19'}$$

It follows that, after the particle has passed the point B', it cannot cross the surface $r = r_2$, but reaches it asymptotically for $\tilde{t} \to \infty$. The same conclusion is drawn from the orientation of the light cone at $r = r_2$, as shown on Figure 24. The motion of the particle as described in the coordinates $\tilde{t}$, r will therefore have the form shown on Figure 26. The fact that the particle arrives at A' after having started from the surface $r = r_1$ at $\tilde{t} = -\infty$ follows from the remark that no particle can cross the surface $r = r_1$ outwards.

It is easy to calculate the proper time interval for the motion of the particle from B' to the point B'' (Figure 26). One finds then that the particle reaches the point B_∞ with coordinates $r = r_2$, $\tilde{t} = +\infty$ in a *finite* proper time interval. I.e. the space described by the coordinates $\tilde{t}$, r is incomplete for the time-like as well as null geodesics passing through B_∞. Similarly one proves that the space is incomplete also for geodesics passing through the point A_∞ with coordinates $r = r_1$, $\tilde{t} = -\infty$.

The extension to a complete manifold can be obtained by the use of Kruskal-type coordinates, but again the so extended manifold has features which are extremely puzzling from the physical point of view.

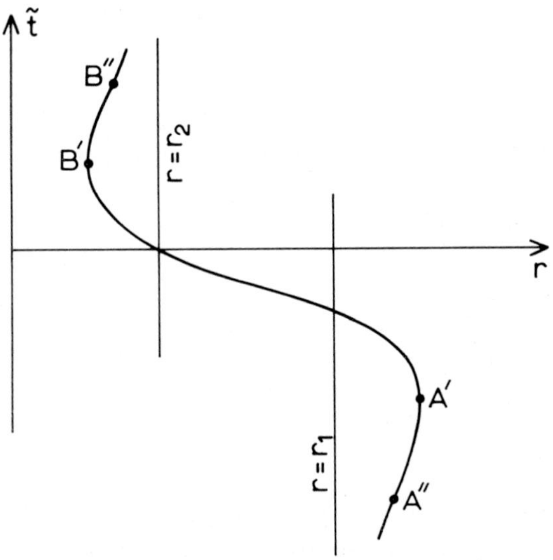

Fig. 26. The motion of a neutral test particle in the diagram $(\tilde{t}, r)$.

39. Some Remarks on 'Unified Field Theories'

The field equations (36.24) and (36.25) for the combined gravitational and electromagnetic field are certainly satisfactory from the macrophysical point of view. It is essentially more difficult to form an opinion about the applicability of this theory to microphysical problems.

A problem of fundamental importance in microphysics is the understanding of the internal structure of elementary particles. A first discussion of this problem was made possible at the beginning of the 20th century by the development of the Maxwell theory. Indeed, one had hoped to explain the inertial masses of the particles by reducing them to the energy of their electromagnetic fields. The subsequent discussion has shown that this hope was not justified, the reason being intuitively a very simple one: The presence of an electric charge of one sign leads to repulsive forces making the formation of a stable particle impossible.

When one takes into account also the gravitational field, this difficulty ceases to exist, since the gravitational force is attractive. The theory of the combined gravitational and electromagnetic field may therefore be useful for the discussion of this problem. We thus see the necessity of establishing a theory of these fields, which would be correct also in microphysics. In the opinion of Einstein this should be a *geometrically unified theory*, obtained by the appropriate new generalization of the geometrical concepts. This should at the same time remove the asymmetric treatment of the two fields in the formulation of Section 36. Indeed, in that formulation the gravitational field has been geometrized, while the electromagnetic field is still described in the same way as in special relativity.

Much work has been done on this problem in the past. We shall give here a brief description of the ideas underlying the two attempts which have been discussed in more detail.

The *5-dimensional theories* are based on the use of a 5-dimensional Riemannian space. In the initial form of this theory, the *Kaluzza-Klein theory*, the fifth coordinate was assigned a restricted special role. Accordingly the components of the metrical tensor $g_{\lambda\mu}$ $(\lambda, \mu = 1, 2, \ldots, 5)$ split in 3 groups: The 10 components g_{ik} $(i, k = 1, 2, \ldots, 4)$, the 4 components g_{i5} and the component g_{55}. The g_{ik} describe the gravitational field in the same way as in general relativity. The components g_{i5} are equivalent to a 4-dimensional vector and describe the electromagnetic field. The component g_{55} is given a constant value and is in this way eliminated from the field variables. The resulting theory has exactly the same field equations as those derived in Section 36.

In a mathematically improved version the 5 coordinates are considered as *projective coordinates* of the physical 4-dimensional space. In the most general case this theory contains three different fields: A (symmetric) *tensor* field, describing the gravitational interaction, a *vector* field corresponding to electromagnetism and a *scalar* field which does not correspond to an already known macroscopic interaction. The special case of the *tensor-scalar* theory, with the vector field suppressed, has been discussed frequently in recent years as an eventual generalization of the Einstein theory of gravitation, i.e. of general relativity.

An entirely different theory was proposed in 1946 by Einstein and Strauss and developed independently by Schrödinger. This theory is based on the use of a *non-riemannian* 4-dimensional space. The basic elements of the theory are a non-symmetric tensor $g_{\lambda\mu}$ and a non-symmetric connection $\Gamma^{\lambda}_{\mu\nu}$, satisfying the relation:

$$g_{\lambda\mu,\nu} - \Gamma^{\alpha}_{\lambda\nu} g_{\alpha\mu} - \Gamma^{\alpha}_{\nu\mu} g_{\lambda\alpha} = 0\,.$$

The proposed interpretation of $g_{\lambda\mu}$ was the following: The symmetric part $g_{(\lambda\mu)}$ should describe the gravitational field as in general relativity, while the antisymmetric part $g_{[\lambda\mu]}$ should represent the electromagnetic field in suitable geometrical units. The detailed discussion of the theory led to several difficulties and it is now almost certain that this theory cannot contain a description of the electromagnetic field.

A possibility which remains open is that this theory may allow a deeper description of the gravitational field in the following sense: The existence of the spin of elementary particles may require the use of a non-symmetric matter tensor and correspondingly of a non-symmetric gravitational potential $g_{\lambda\mu}$. This theory would then be a rival to the Einstein-Cartan theory, which uses a symmetric $g_{\lambda\mu}$ coupled to a non-symmetric connection. The Einstein-Schrödinger theory may be better adapted to the problem, but it is much more complicated than the Einstein-Cartan theory.

Exercise

IX1: Prove that the equation

$$r = \frac{r_1 + r_2}{2} + \frac{r_1 - r_2}{2} \cos(\theta + \lambda), \qquad \lambda = \text{const}$$

determines a characteristic hypersurface Σ of the Reissner-Nordström metric. Discuss the position of Σ relative to the characteristic surfaces defined by the equations $r = r_1$ and $r = r_2$.

CHAPTER X

EQUATIONS OF MOTION IN GENERAL RELATIVITY

40. The Problem of the Equations of Motion

A field theory has to describe quantitatively a certain type of interaction between material bodies. The field equations alone will in general not be sufficient for this purpose. Besides the field equations we have to postulate also the equations of motion of the material bodies under the influence of the interaction, i.e. we have to introduce a formula for the force density representing the action of the field on each body.

As a first example we mention the Newtonian theory of gravitation. Besides the field equation (17.1),

$$\Delta\varphi = -4\pi G\varrho\,,$$

we have to postulate that the (3-dimensional) force density is given by

$$f_i = -\varrho\frac{\partial\varphi}{\partial x^i}\,. \tag{40.1}$$

A similar situation presents itself in the case of the electromagnetic field. Besides the Maxwell equations we have to postulate the (4-dimensional) force density

$$f_\lambda = F_{\lambda\mu}s^\mu\,. \tag{40.2}$$

In general relativity the geometrization of the gravitational field makes it impossible to introduce a force density having tensor properties. This follows immediately from the principle of equivalence, according to which gravitational forces are of the same nature as inertial forces. Consequently it was necessary to postulate, besides the field equations, an equation of motion of a different type. The fact that the gravitational field was described geometrically led then directly to the postulate of geodesic motion.

Later it was realized that the geodesic postulate could be used only for the motion of *test particles*. These are particles or small bodies moving in the neighbourhood of much more massive bodies. Because of the smallness of the mass of the test particle its contribution to the total field will also be small. Consequently we can neglect this contribution and still have a sufficiently good approximation. We then talk of the *background metric*, produced by the massive bodies in the absence of the test particle. The law of geodesic motion is then to be so understood, that the test particle will move on a geodesic of this background metric.

In the case of bodies of comparable masses the gravitational field varies appreciably in the interior of each one of the bodies. It follows that it will then be impossible to define a background field and consequently the geodesic postulate cannot be applied in an unambiguous way. Actually the discussion of this problem in more detail led to an important discovery: *In general relativity the equations of motion of material bodies are determined by the field equations.*

This means that in general relativity we cannot proceed as for example in the Maxwell theory, where we completed the field equations by a postulate leading to the equations of motion. Instead we have in general relativity the new problem of deriving the equations of motion from the field equations. This is the problem which will be discussed briefly in this chapter.

It is not difficult to explain qualitatively why in general relativity the equations of motion are determined by the field equations. We start with a remark on the Maxwell theory. The expression (40.2) for the electromagnetic force density can be derived from the stress-energy-momentum tensor of the electromagnetic field, as is shown by Equation (36.12). It follows that the equations of motion can be derived from the differential form of the conservation law of energy and momentum, which in this case is given by (36.5):

$$\eta^{\mu\nu}(T_{\lambda\mu} + {}_{\mathrm{M}}T_{\lambda\mu})_{,\nu} = 0. \tag{40.3}$$

We thus see that in special relativity the differential form of the conservation law of energy and momentum plays the role of the *dynamical equation*, in the sense that one can derive from it the equations of motion of particles when their internal structure is given.

In general relativity it is the covariant equation (19.5)

$$T^{\lambda\mu}{}_{;\mu} = 0 \tag{40.4}$$

which expresses the conservation law of energy and momentum in the presence of the gravitational interaction. It is therefore this equation which will play now the role of the dynamical equation.

The important new feature which we encounter in general relativity is that Equation (40.4) is a consequence of the field equations (19.7):

$$R^{\lambda\mu} - \tfrac{1}{2}g^{\lambda\mu}R = -\kappa T^{\lambda\mu},$$

because of the identity (19.6) satisfied by the left-hand side of this equation:

$$(R^{\lambda\mu} - \tfrac{1}{2}g^{\lambda\mu}R)_{;\mu} = 0.$$

This is the reason why in general relativity the equations of motion are determined by the field equations.

Evidently we shall have the same situation in any field theory in which the conservation law of energy and momentum is a consequence of the field equations. In particular we shall have the same situation in the Einstein-Maxwell theory, since in this theory the conservation law (36.10) is again a consequence of the field equations (36.9).

The problem of deriving the equations of motion from the field equations of general relativity is very difficult from the mathematical point of view. A number of special cases have been discussed, always using approximation methods. The following two cases will be examined in the next two sections.

(1) *The 'astronomical' problem*, dealing with the motion of a system of bodies of comparable masses, as e.g. a double star. This case is characterized by the following assumptions:

(a) The gravitational field is weak, at least outside the bodies, and

(b) The motions of all bodies are non-relativistic, i.e. the velocity v of any one of the bodies satisfies the condition

$$v/c \ll 1\,. \tag{40.5}$$

The calculations needed for the solution of this problem are so long that it will be impossible to discuss it here quantitatively. We shall limit ourselves to a brief description of the procedure leading to the solution and of the results which have been obtained.

(2) The problem of the *motion of a test particle.* In this case the calculations are essentially simpler and a limitation of the form (40.5) is not needed. We shall examine in detail in Section 42 the case of the simplest type of test particle which is the *single-pole test particle.*

41. The 'Astronomical' Problem

This is the first problem of motion which was discussed systematically. Two different methods were developed for its discussion almost simultaneously. Common to these methods is the expansion of $g_{\lambda\mu}$ in power series of a small parameter, which is formally the quantity $1/c$. (In reality it is a quantity of the order v/c.)

The first method, the Einstein-Infeld-Hoffman method (abbreviated as the EIH-method), uses only the field equations valid in the empty space surrounding the bodies. The field equations are split into the equations of different orders $1/c^n$ $(n=1, 2\ldots)$, the integration starting from the equations of the lowest order. It is found that for the equations of order $n+1$ there are integrability conditions containing the field quantities of orders $m \leqslant n$. These integrability conditions are expressed in the form of integrals over 2-dimensional surfaces surrounding each one of the bodies. Actually the integrability conditions for the equations of order $n+1$ determine the equations of motion of the bodies to the order n.

The result which was obtained to order $n=2$ is identical with the Newtonian equations of motion. After some extremely long calculations it was also possible to derive the equations of motion to order $n=4$. These equations determine relativistic corrections, one of which is a generalization of the advance of the perihelion effect.

This method has the advantage of being applicable to the case of bodies having arbitrarily small dimensions. Because of the extremely long calculations, which are needed in this method, only the case of bodies, which are spherically symmetric in their restframe, has been discussed.

In the second method, the Fock method, one considers also the interior of the bodies, which is described by a matter tensor of the simple form of perfect fluid (Section 28).

The gravitational field is now assumed to be weak everywhere, i.e. outside as well as inside the bodies, this assumption leading necessarily to a lower limit of the dimensions of the bodies. The integrability conditions are found to be related to the conservation law (40.4) and are now expressed by 3-dimensional integrals of Equation (40.4) over the interior of each one of the bodies. The fact that this method makes a systematic use of the equation (40.4) has the consequence that the calculations needed for the derivation of the equations of motion are much simpler than those needed in the EIH-method.

The results obtained by this second method are of course identical, for spherically symmetric bodies, with those obtained by the EIH-method. Using the Fock method it was possible to discuss also the motion of bodies having angular momentum.

More recently this method has been developed further and used systematically for the purpose of discussing several questions related to the internal structure of astronomical bodies.

42. The Motion of Test Particles

As we mentioned already in Section 40, a test particle is a body of small mass moving in the neighbourhood of much more massive bodies. It follows at once that we can obtain a useful approximation method if we start from the metric ${}_0g_{\lambda\mu}$ describing the gravitational field of the massive bodies in the absence of the test particle. We shall assume that the metric ${}_0g_{\lambda\mu}$ is known.

When the test particle is present, the metric $g_{\lambda\mu}$ will be:

$$g_{\lambda\mu} = {}_0g_{\lambda\mu} + \delta g_{\lambda\mu}, \tag{42.1}$$

the small correction $\delta g_{\lambda\mu}$ being caused by the test particle. In the first approximation it will be sufficient to retain in the field equations only the terms which are linear in $\delta g_{\lambda\mu}$.

This problem also has been treated by two different methods. The first method is analogous to the EIH-method used for the discussion of the astronomical problem. One considers only the exterior of the test particle (and of the massive bodies). One starts by deriving the field equations of the first order in $\delta g_{\lambda\mu}$ and then one establishes the integrability conditions, in the form of integrals over a 2-dimensional surface surrounding the test particle. These integrability conditions lead to the equations of motion of the test particle.

In the second method one considers also the interior of the test particle. Let the material distribution inside the particle be described by the matter tensor $\delta T^{\lambda\mu}$. We shall assume that the test particle is moving outside the massive bodies, i.e. that the matter tensor ${}_0T^{\lambda\mu}$ corresponding to the metric ${}_0g_{\lambda\mu}$ vanishes inside as well as near the test particle. Consequently the tensor $\delta T^{\lambda\mu}$ will satisfy the conservation law (19.5):

$$(\delta T^{\lambda\mu})_{;\mu} = 0. \tag{42.2}$$

We shall now prove that in this second method it is not necessary to consider the linearized field equations for $\delta g_{\lambda\mu}$: The use of the conservation law (42.2) alone is sufficient to derive the equations of motion of the test particle.

In the 4-dimensional diagram let the interior of the particle be the tube-like region contained in the time-like hypersurface Σ (Figure 27). The matter tensor $\delta T^{\lambda\mu}$ describing the interior of the particle will be non-zero inside Σ and zero outside. It is not necessary in this problem to assume that $\delta T^{\lambda\mu}$ is of any particular form.

It will be more convenient to use Equation (42.2) multiplied be $\sqrt{-g}$. Writing

$$\sqrt{-g}\,\delta T^{\lambda\mu} \equiv \mathfrak{T}^{\lambda\mu} \tag{42.3}$$

and using the formula (6.20) we find the relation:

$$\mathfrak{T}^{\lambda\mu}{}_{,\mu} + \Gamma^{\lambda}_{\mu\nu}\mathfrak{T}^{\mu\nu} = 0\,. \tag{42.4}$$

The Christoffel symbols $\Gamma^{\lambda}_{\mu\nu}$ can be written in the form:

$$\Gamma^{\lambda}_{\mu\nu} = {}_0\Gamma^{\lambda}_{\mu\nu} + \delta\Gamma^{\lambda}_{\mu\nu}\,, \tag{42.5}$$

${}_0\Gamma^{\lambda}_{\mu\nu}$ being the values corresponding to the metric ${}_0g_{\lambda\mu}$ and $\delta\Gamma^{\lambda}_{\mu\nu}$ the small corrections because of the term $\delta g_{\lambda\mu}$ in (42.1). The approximation of the test particle consists in neglecting the term $\delta\Gamma^{\lambda}_{\mu\nu}$ when we introduce (42.5) in (42.4). Thus we get the equation

$$\mathfrak{T}^{\lambda\mu}{}_{,\mu} + {}_0\Gamma^{\lambda}_{\mu\nu}\mathfrak{T}^{\mu\nu} = 0\,, \tag{42.6}$$

which is the basis of the subsequent calculations.

The next assumption is that the test particle has small linear dimensions. In a more exact manner we assume that the linear dimensions of the cross-section of the tube Σ (Figure 28) are small compared with a length R characterizing the gradient of the metric ${}_0g_{\lambda\mu}$ (for example the distance of the test particle from the massive bodies).

We shall choose some smooth curve inside the tube Σ (Figure 27). Let this 'axis' of

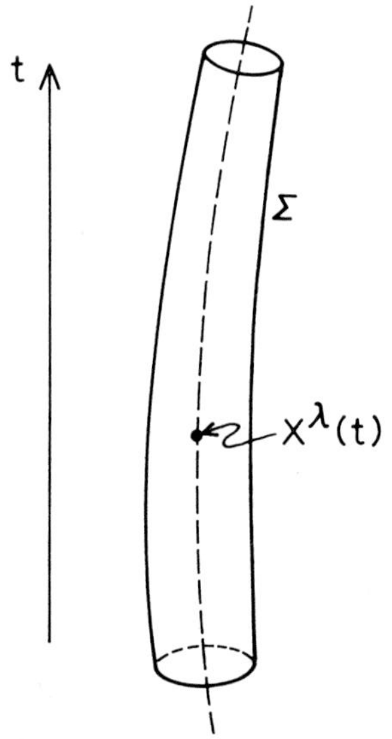

Fig. 27. The hypersurface Σ is the boundary of the interior of the test particle.

the tube define in the different sections $t=\text{const}$ of the tube the points with coordinates $X^\lambda(t)$. Since ${}_0g_{\lambda\mu}$ changes very little inside the section of the tube, we can develop ${}_0\Gamma^\lambda_{\mu\nu}(x^\varrho)$ in a Taylor series around the point X^λ (Figure 28):

$$
{}_0\Gamma^\lambda_{\mu\nu}(x^\varrho) = {}_0\Gamma^\lambda_{\mu\nu}(X^\varrho) + (x^\sigma - X^\sigma)\frac{\partial_0\Gamma^\lambda_{\mu\nu}(X^\varrho)}{\partial X^\sigma} + \cdots. \tag{42.7}
$$

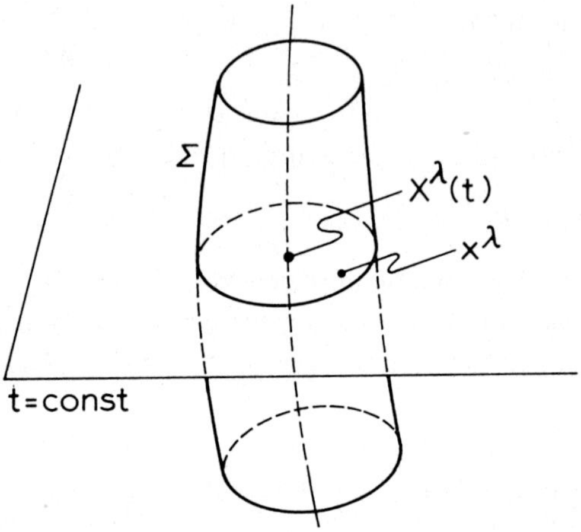

Fig. 28. A section of the interior of the test particle by a hypersurface $t=\text{const}$

We shall consider here the test particle with the simplest internal structure, which is the structure of a *single-pole*. Having in mind the limiting case of a point-like particle we shall define the single-pole particle by the requirement that its dipole as well as higher multipole moments vanish:

$$
\int (x^\varrho - X^\varrho)\,\mathfrak{T}^{\lambda\mu}\,\mathrm{d}^3x = 0, \quad \int (x^\varrho - X^\varrho)(x^\sigma - X^\sigma)\,\mathfrak{T}^{\lambda\mu}\,\mathrm{d}^3x = 0, \ldots \tag{42.8}
$$

In reality we only assume that the integrals (42.8) are negligible, since we are considering particles which are not strictly point-like.

We now introduce (42.7) into (42.6) and we integrate the resulting equation over the hypersurface $t=\text{const}$. Remembering that $\mathfrak{T}^{\lambda\mu}$ vanishes outside Σ and taking into account the conditions (42.8) for a single-pole particle we find:

$$
\frac{\mathrm{d}}{\mathrm{d}t}\int \mathfrak{T}^{\lambda 0}\,\mathrm{d}^3x + {}_0\Gamma^\lambda_{\mu\nu}(X^\varrho)\int \mathfrak{T}^{\mu\nu}\,\mathrm{d}^3x = 0. \tag{42.9}
$$

The following equation is an immediate consequence of (42.6):

$$
\mathfrak{T}^{\mu\nu} = (x^\mu \mathfrak{T}^{\varrho\nu})_{,\varrho} + x^\mu{}_0\Gamma^\nu_{\varrho\sigma}\mathfrak{T}^{\varrho\sigma}. \tag{42.10}
$$

Integrating this equation over the hypersurface $t = \text{const}$ we find:

$$\int \mathfrak{T}^{\mu\nu}\,\mathrm{d}^3x = \frac{\mathrm{d}}{\mathrm{d}t}\int x^\mu \mathfrak{T}^{\nu 0}\,\mathrm{d}^3x + {}_0\Gamma^\nu_{\varrho\sigma}(X^\alpha)\int x^\mu \mathfrak{T}^{\varrho\sigma}\,\mathrm{d}^3x. \tag{42.11}$$

Because of (42.8) we can write:

$$\left.\begin{aligned} \int x^\mu \mathfrak{T}^{\nu 0}\,\mathrm{d}^3x &= X^\mu(t)\int \mathfrak{T}^{\nu 0}\,\mathrm{d}^3x;\\ \int x^\mu \mathfrak{T}^{\varrho\sigma}\,\mathrm{d}^3x &= X^\mu(t)\int \mathfrak{T}^{\varrho\sigma}\,\mathrm{d}^3x. \end{aligned}\right\} \tag{42.12}$$

Introducing (42.12) into (42.11) and taking into account (42.9) we find:

$$\int \mathfrak{T}^{\mu\nu}\,\mathrm{d}^3x = \frac{\mathrm{d}X^\mu}{\mathrm{d}t}\int \mathfrak{T}^{\nu 0}\,\mathrm{d}^3x. \tag{42.13}$$

From (42.13) we find, for $\nu = 0$:

$$\int \mathfrak{T}^{\mu 0}\,\mathrm{d}^3x = \frac{\mathrm{d}X^\mu}{\mathrm{d}t}\int \mathfrak{T}^{00}\,\mathrm{d}^3x. \tag{42.14}$$

Introducing this in (42.13) we obtain the more general result:

$$\int \mathfrak{T}^{\mu\nu}\,\mathrm{d}^3x = \frac{\mathrm{d}X^\mu}{\mathrm{d}t}\frac{\mathrm{d}X^\nu}{\mathrm{d}t}\int \mathfrak{T}^{00}\,\mathrm{d}^3x. \tag{42.15}$$

Introducing (42.15) into Equation (42.9) we find:

$$\frac{\mathrm{d}}{\mathrm{d}t}\left(\frac{\mathrm{d}X^\lambda}{\mathrm{d}t}\int \mathfrak{T}^{00}\,\mathrm{d}^3x\right) + {}_0\Gamma^\lambda_{\mu\nu}\frac{\mathrm{d}X^\mu}{\mathrm{d}t}\frac{\mathrm{d}X^\nu}{\mathrm{d}t}\int \mathfrak{T}^{00}\,\mathrm{d}^3x = 0. \tag{42.16}$$

This is already the equation of motion of the particle, which however has to be brought into a more convenient form. We introduce the line-element of the curve $X^\lambda(t)$ and the corresponding 4-dimensional velocity:

$$\mathrm{d}s^2 = {}_0g_{\lambda\mu}\,\mathrm{d}X^\lambda\,\mathrm{d}X^\mu; \qquad u^\lambda = \frac{\mathrm{d}X^\lambda}{\mathrm{d}s}. \tag{42.17}$$

Moreover we put:

$$\frac{1}{u^0}\int \mathfrak{T}^{00}\,\mathrm{d}^3x = m_0 \qquad \left(u^0 = \frac{\mathrm{d}t}{\mathrm{d}s}\right). \tag{42.18}$$

Equation (42.16) takes then the form:

$$\frac{\mathrm{d}}{\mathrm{d}s}(m_0 u^\lambda) + m_0\,{}_0\Gamma^\lambda_{\mu\nu}u^\mu u^\nu = 0. \tag{42.19}$$

The equation of motion (42.19) can be split into two simpler equations. We develop the first term,

$$\frac{\mathrm{d}}{\mathrm{d}s}(m_0 u^\lambda) = m_0 \frac{\mathrm{d}u^\lambda}{\mathrm{d}s} + \frac{\mathrm{d}m_0}{\mathrm{d}s} u^\lambda,$$

and then we multiply Equation (42.19) by u_λ. Since:

$$u_\lambda \left(\frac{\mathrm{d}u^\lambda}{\mathrm{d}s} + {}_0\Gamma^\lambda_{\mu\nu} u^\mu u^\nu \right) = u_\lambda \, u^\lambda{}_{;\,\mu} \, u^\mu = 0,$$

because of (13.9), we find:

$$\frac{\mathrm{d}m_0}{\mathrm{d}s} = 0. \tag{42.20}$$

Equation (42.19) reduces then to:

$$\frac{\mathrm{d}u^\lambda}{\mathrm{d}s} + {}_0\Gamma^\lambda_{\mu\nu} u^\mu u^\nu \equiv u^\lambda{}_{;\,\mu} \, u^\mu = 0. \tag{42.21}$$

Equation (42.21) is the geodesic equation. We thus see that the detailed discussion of the problem of motion confirms the geodesic postulate for the single-pole test particle. Additionally we found the result (42.20) stating that the rest-mass of the particle is a constant of the motion.

We mention, without entering into the details, that this method has been used also to derive the equations of motion of a test particle with angular momentum or, in other words, a *spinning* test particle. The feature which characterizes this case is the non-vanishing of at least one of the dipole moments

$$M^{\lambda\mu 0} = \int (x^\lambda - X^\lambda) \, \mathfrak{T}^{\mu 0} \, \mathrm{d}^3 x .$$

One obtains in this case equations of motion for the position $X^\lambda(t)$ of the particle and for its angular momentum or *spin* $J^{\lambda\mu}$,

$$J^{\lambda\mu} = M^{\lambda\mu 0} - M^{\mu\lambda 0} .$$

In general the trajectory of the spinning test particle is not a geodesic.

We also mention that the calculations can be made simpler and more elegant if one considers the particle as point-like, in which case $\mathfrak{T}^{\lambda\mu}$ will be represented by δ-functions. There is, however, a certain incompatibility in this case. Indeed, with the mass of the test particle being small but certainly different from zero, the correction $\delta g_{\lambda\mu}$ would be infinitely large at the position of a point-like particle. In this respect the first method, which considers the field only outside the particle, is really advantageous, since it can be used also for test particles which are strictly point-like.

We close this section with the following remark. In the approximation considered here any gravitational radiation emitted by the test particle and any effect of this radiation on the motion of the particle has been neglected. As we know from the discussion of the similar problem in the electromagnetic theory, these effects can be taken into account in a higher approximation, in which it will be necessary to determine correctly the contribution $\delta g_{\lambda\mu}$ of the test particle to the total field.

Exercise

X1: Derive the equations of motion of a charged test particle having the structure of a simple pole with respect to the mass as well as to the electric charge.

CHAPTER XI

GRAVITATIONAL RADIATION

43. Introductory Remarks

The concept of *radiation* has emerged at the same time as the concept of *field* from the quantitative study of electromagnetic phenomena as described by the Maxwell theory.

The basically new feature of this theory is that the Maxwell equations are differential equations of the hyperbolic type. This has the consequence that changes of the electromagnetic forces caused by disturbances in the distribution of the electromagnetic sources, as is for example the accelerated motion of a charged body, are propagated through space with a *finite speed.* Consequently the action of one charged body on another is not simultaneous but *retarded.*

The discussion of the conservation law of energy and momentum has shown that because of this retardation effect it is necessary to introduce a distribution of *electromagnetic* energy and momentum of the electromagnetic field.

Actually hyperbolic field equations have not only retarded, but also advanced solutions. The discussion of the requirements of the causality principle has shown that we have to use the retarded solution only. In the case of the Maxwell equations, which are linear differential equations with constant coefficients, we have explicit formulae determining the retarded solutions when the sources are given. With the help of these formulae it is found that sources in accelerated motion create a field containing a distribution of electromagnetic energy and momentum which *moves away from the sources.* This is the *electromagnetic radiation* emitted by such sources.

The Einstein equations of the gravitational field are also hyperbolic, as we shall see in detail later. Therefore we have in general relativity a situation similar to that found in Maxwell theory. In particular general relativity leads necessarily to the existence of *gravitational radiation.*

Because of the non-linearity of the Einstein equations, the detailed study of gravitational radiation is a very difficult problem. Until now only some special aspects of this problem have been discussed and certain partial results obtained. We shall describe in this chapter the more important of these results. They can be divided in the following three parts:

(1) Gravitational radiation in the linear approximation. It will be examined in Sections 44–46.

(2) Gravitational shock waves will be treated exactly in Sections 47 and 48.

(3) The 'asymptotically exact' treatment of the propagation of gravitational radiation, at large distance from the sources, will be described in Sections 49–51.

44. The Linear Approximation

This approximation has been discussed already in Section 31. The basic assumption is that the gravitational field is weak everywhere, i.e. outside as well as inside the bodies which are the sources of the field. It will be more convenient to use systematically as field variables the components of the contravariant tensor density

$$\mathfrak{g}^{\lambda\mu} = \sqrt{-g}\, g^{\lambda\mu}.$$

We write, instead of (31.3):

$$\mathfrak{g}^{\lambda\mu} = \eta^{\lambda\mu} + \mathfrak{g}'^{\lambda\mu} + \mathfrak{g}''^{\lambda\mu} + \cdots. \tag{44.1}$$

In the linear approximation we retain only the term $\mathfrak{g}'^{\lambda\mu}$ of first order. To simplify the subsequent formulae we shall write:

$$\mathfrak{g}'^{\lambda\mu} \equiv \upsilon^{\lambda\mu}. \tag{44.2}$$

One can verify without difficulty the following relation between $\upsilon^{\lambda\mu}$ and the quantity $\gamma_{\lambda\mu}$ used in Section 31:

$$\upsilon^{\lambda\mu} = -\,\eta^{\lambda\varrho}\eta^{\mu\sigma}\gamma_{\varrho\sigma}. \tag{44.3}$$

It follows that the coordinate condition (31.21) can be written in the form:

$$\upsilon^{\lambda\mu}{}_{,\mu} = 0. \tag{44.4}$$

We mention that this is the *harmonic* coordinate condition:

$$\mathfrak{g}^{\lambda\mu}{}_{,\mu} = 0 \tag{44.5}$$

applied to the first-order term $\mathfrak{g}'^{\lambda\mu}$. The coordinate condition (44.5) has been found useful in the discussion of several problems.

Using the quantity $\upsilon^{\lambda\mu}$ we write the linearized field equations (31.22) in the form:

$$\Box\, \upsilon^{\lambda\mu} = 2\kappa\, {}_0\mathfrak{T}^{\lambda\mu}. \tag{44.6}$$

The quantity ${}_0\mathfrak{T}^{\lambda\mu}$ is the term of order zero in the development of the tensor density

$$\mathfrak{T}^{\lambda\mu} \equiv \sqrt{-g}\, T^{\lambda\mu}.$$

Taking into account the relation

$$g = -1 + \cdots,$$

the dots being terms of the first order, we find at once:

$${}_0\mathfrak{T}^{\lambda\mu} = \eta^{\lambda\varrho}\eta^{\mu\sigma}\, {}_0T_{\varrho\sigma};$$

${}_0T_{\varrho\sigma}$ is the quantity appearing in Equation (31.22). The conservation law (31.14) takes now the form:

$${}_0\mathfrak{T}^{\lambda\mu}{}_{,\mu} = 0. \tag{44.7}$$

Equation (44.7) is the conservation law of energy and momentum in special relativity. The fact that the exact conservation law (19.5), which is equivalent to

$$\mathfrak{T}^{\lambda\mu}{}_{;\mu} = 0\,,$$

has been reduced to the form (44.7) is a consequence of the assumption that the gravitational field is weak: In the linear approximation the gravitational force is assumed to be negligible compared with the non-gravitational forces. It follows that the results of the subsequent calculations will be applicable only to cases in which non-gravitational forces are predominant, as is for example the case of a vibrating elastic body. When the non-gravitational forces are comparable to the gravitational ones or negligible, the calculation has to be continued in the second approximation. This is a much more difficult problem, which has not yet been solved in a satisfactory way.

Two different methods have been used for studying gravitational radiation in the linear approximation. In the first method one considers the field far from its sources and the final results are expressed by integrals over a 2-dimensional sphere of radius $r \to \infty$. In the second method one considers the interior of the sources and the final results are given by integrals over the sources. We shall describe here only the first method.

The basic relation in this method is the pseudotensorial law of conservation of energy and momentum. We shall use the form (35.5) of this pseudotensorial equation:

$$(\mathfrak{T}_\lambda{}^\mu + \mathfrak{t}_\lambda{}^\mu)_{,\mu} = 0\,. \tag{44.8}$$

The pseudotensor $\mathfrak{t}_\lambda{}^\mu$ is given by Equation (34.25), using the expressions (33.20) and (33.22). The lowest order term $\mathfrak{t}'_\lambda{}^\mu$ of $\mathfrak{t}_\lambda{}^\mu$ will be obtained if we replace the derivatives $\mathfrak{g}^{\lambda\mu}{}_{,\nu}$ by $\upsilon^{\lambda\mu}{}_{,\nu}$ and the non-differentiated factors $\mathfrak{g}^{\lambda\mu}$ by $\eta^{\lambda\mu}$. We thus find that the lowest order term $\mathfrak{t}'_\lambda{}^\mu$ is given by the following lengthy expression:

$$\left.\begin{aligned} \mathfrak{t}'^{\,\mu}_{\lambda} = \frac{1}{2\kappa}\,\{&\tfrac{1}{2}\eta^{\mu\beta}\left(\eta_{\varrho\nu}\eta_{\alpha\sigma} - \tfrac{1}{2}\eta_{\varrho\sigma}\eta_{\nu\alpha}\right)\upsilon^{\nu\alpha}{}_{,\beta}\upsilon^{\varrho\sigma}{}_{,\lambda} - \\ &- \eta_{\varrho\sigma}\upsilon^{\varrho\beta}{}_{,\lambda}\upsilon^{\mu\sigma}{}_{,\beta} + \tfrac{1}{2}\delta^\mu_\lambda\eta_{\alpha\beta}\upsilon^{\alpha\varrho}{}_{,\sigma}\upsilon^{\beta\sigma}{}_{,\varrho} - \\ &- \tfrac{1}{4}\delta^\mu_\lambda\eta^{\alpha\beta}\left(\eta_{\varrho\nu}\eta_{\sigma\gamma} - \tfrac{1}{2}\eta_{\varrho\sigma}\eta_{\nu\gamma}\right)\upsilon^{\varrho\sigma}{}_{,\alpha}\upsilon^{\nu\gamma}{}_{,\beta}\}\,. \end{aligned}\right\} \tag{44.9}$$

We have already mentioned the difficulties arising from the pseudotensorial character of Equation (44.8). As we explained, however, in Section 35, these difficulties disappear when we use this equation in order to determine the total energy and momentum of a field which is stationary at large distances. This remark will enable us to write a formula for the total energy and momentum of the gravitational radiation in the following case. We consider a field produced by sources which have finite extension in the 3-dimensional space:

$${}_0\mathfrak{T}^{\lambda\mu} = 0 \quad \text{for} \quad r > a\,. \tag{44.10}$$

We assume that the field and its sources are initially stationary. The sources start, at a certain moment, some internal motion, e.g. an elastic vibration. This motion is stopped after a finite time interval. The gravitational field will then become again time-independent. This is shown schematically on Figure 29.

If we denote by R_μ the total energy ($\mu=0$) and momentum ($\mu=i=1, 2, 3$) carried away by the gravitational radiation, we shall have:

$$R_\mu = (P_\mu)_{\text{in}} - (P_\mu)_{\text{fin}}. \tag{44.11}$$

$(P_\mu)_{\text{in}}$ and $(P_\mu)_{\text{fin}}$ are respectively the total energy and momentum of the initial and the final stationary field. These quantities are given correctly by the formula (35.6), derived from (44.8):

$$(P_\mu)_{\text{in}} = \int\limits_{t=t_1} (\mathfrak{T}^0_\mu + \mathfrak{t}^0_\mu)\, \mathrm{d}^3x; \qquad (P_\mu)_{\text{fin}} = \int\limits_{t=t_2} (\mathfrak{T}^0_\mu + \mathfrak{t}^0_\mu)\, \mathrm{d}^3x. \tag{44.12}$$

Let us now integrate Equation (44.8) over the 4-dimensional region which is contained by the hypersurfaces $t=t_1$, Σ and $t=t_2$ (Figure 29). The integral can be transformed to an integral over the boundary and the result is:

$$\int\limits_{t=t_2} (\mathfrak{T}^0_\mu + \mathfrak{t}^0_\mu)\, \mathrm{d}^3x - \int\limits_{t=t_1} (\mathfrak{T}^0_\mu + \mathfrak{t}^0_\mu)\, \mathrm{d}^3x + \int\limits_{\Sigma} (\mathfrak{T}^i_\mu + \mathfrak{t}^i_\mu)\, \mathrm{d}\sigma_i = 0. \tag{44.13}$$

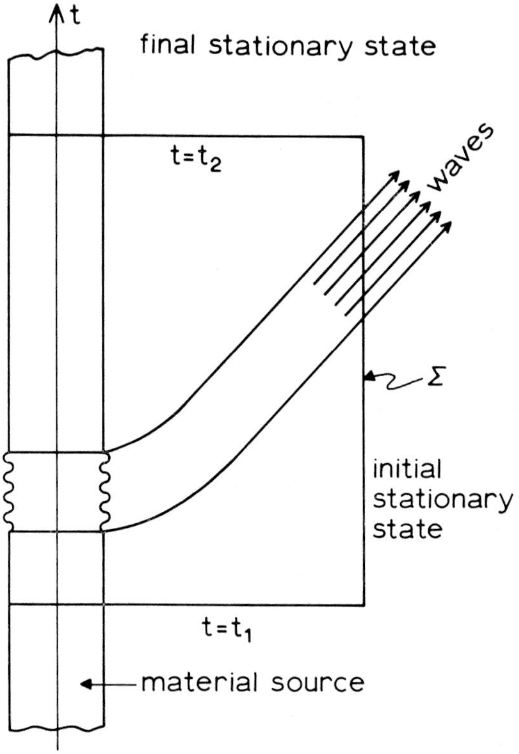

Fig. 29. Schematic diagram showing a field which is stationary initially and finally.

The element $d\sigma_i$ of the hypersurface Σ can be written as follows:

$$d\sigma_i = n_i \, dS \, dt ,$$

where dS is the element of the 2-dimensional sphere S defined by the equation $r = \text{const} \to \infty$ and n_i the normal to S. Since $\mathfrak{T}_\mu^i$ vanishes on Σ according to (44.10), we find from (44.13), combined with (44.11) and (44.12):

$$R_\mu = \int_{t_1}^{t_2} dt \int_S \mathfrak{t}_\mu^i n_i \, dS . \tag{44.14}$$

This result can be written also in the form:

$$R_\mu = \int_{t_1}^{t_2} \frac{dR_\mu}{dt} dt; \qquad \frac{dR_\mu}{dt} = \int_S \mathfrak{t}_\mu^i n_i \, dS . \tag{44.15}$$

In the linear approximation we have to write $\mathfrak{t}_\mu'^i$ instead of $\mathfrak{t}_\mu^i$.

The following remarks should be made here. Because of the complicated structure of the gravitational field equations there is, in the propagation of gravitational radiation, the effect of back-scattering. This has the consequence that the final stationary field will be obtained, speaking strictly, only asymptotically for $t \to +\infty$. However, this does not invalidate the reasoning leading to Equation (44.14).

The second remark is that if the total momentum of the radiation does not vanish, $R_i \neq 0$, the final field will be stationary in some other coordinate system. The Figure 29 corresponds to the case $R_i = 0$.

Finally we recall that, as we have seen in Section 35, there are several forms of the pseudotensorial conservation law of energy and momentum, which however give the same total values for the energy and momentum of a stationary field. Since the reasoning which led to (44.14) is based on Equation (44.11), we conclude that we can rewrite Equation (44.14) with any one of these pseudotensors and still obtain the correct total value of the energy and momentum of the radiation. The following calculations will be based on the Einstein pseudotensor $\mathfrak{t}_\lambda{}^\mu$, given to first order by Equation (44.9).

45. The Gravitational Poynting Vector

In the integral (44.14) the surface element dS is proportional to r^2. Since the sphere S of integration has a radius $r \to \infty$, we can neglect all terms in $\mathfrak{t}_\mu'^i$ which vanish at $r \to \infty$ faster than r^{-2}. The expression for $\mathfrak{t}_\mu'^i$ being quadratic in the derivatives $v^{\lambda\mu}{}_{,\nu}$, we see at once that we can neglect in $v^{\lambda\mu}{}_{,\nu}$ all terms which vanish faster than r^{-1}.

The retarded solution of Equation (44.6) is, according to (31.24):

$$v^{\lambda\mu}(x^i, t) = \frac{\kappa}{2\pi} \int \frac{1}{R} \, {}_0\mathfrak{T}^{\lambda\mu}(X^i, T) \, d^3X . \tag{45.1}$$

T is the retarded time:

$$T = t - R. \tag{45.2}$$

The distance R is given by Equation (31.25):

$$R^2 = r^2 - 2x^iX^i + X^iX^i, \qquad r^2 = x^ix^i. \tag{45.3}$$

The quantities $v^{\lambda\mu}$ are needed in (44.14) at distances $r \to \infty$. For $r \gg |X^i|$ we derive immediately from (45.3):

$$R = r + O(1); \qquad \frac{1}{R} = \frac{1}{r} + O\left(\frac{1}{r^2}\right). \tag{45.4}$$

Therefore from (45.1):

$$v^{\lambda\mu}(x^i, t) = \frac{\kappa}{2\pi r} \int {}_0\mathfrak{T}^{\lambda\mu}(X^i, T)\, \mathrm{d}^3X + O\left(\frac{1}{r^2}\right). \tag{45.5}$$

In the derivative $v^{\lambda\mu}{}_{,\nu}$ the terms of order $1/r$ are those arising from the derivation of the retarded time T:

$$v^{\lambda\mu}{}_{,\nu} = \frac{\kappa}{2\pi r} \left\{ \int \frac{\partial}{\partial T}\, {}_0\mathfrak{T}^{\lambda\mu}\, \mathrm{d}^3X \right\} \frac{\partial T}{\partial x^\nu} + O\left(\frac{1}{r^2}\right).$$

From (45.2) and (45.3) we find:

$$\frac{\partial T}{\partial x^0} = \frac{\partial T}{\partial t} = 1, \qquad \frac{\partial T}{\partial x^i} = -\frac{\partial R}{\partial x^i} = -\frac{x^i}{r} + O\left(\frac{1}{r}\right).$$

Therefore:

$$v^{\lambda\mu}{}_{,\nu} = \frac{\kappa}{2\pi r}\, \Omega^{\lambda\mu}\xi_\nu + O\left(\frac{1}{r^2}\right); \tag{45.6}$$

$$\Omega^{\lambda\mu} = \int \frac{\partial}{\partial T}\, {}_0\mathfrak{T}^{\lambda\mu}\, \mathrm{d}^3X, \tag{45.7}$$

$$\xi_\nu \equiv \left(1, -\frac{x^i}{r}\right) \quad \Leftrightarrow \quad \xi^\nu = \eta^{\nu\varrho}\xi_\varrho = \left(1, \frac{x^i}{r}\right). \tag{45.8}$$

The vector ξ^ν is isotropic with respect to the Minkowski metric $\eta_{\lambda\mu}$:

$$\eta_{\lambda\mu}\xi^\lambda\xi^\mu = 0, \tag{45.9}$$

and determines, in the linear approximation, the direction of propagation of the radiation at $r \to \infty$.

The 10 quantities $\Omega^{\lambda\mu}$ are not independent. Indeed, when we retain only terms of order $1/r$, we derive from (44.4) the relation:

$$\Omega^{\lambda\mu}\xi_\mu = 0. \tag{45.10}$$

Consequently only the 6 components Ω^{ik} ($i, k=1, 2, 3$) are independent. The remaining 4 quantities Ω^{0i} and Ω^{00} are given by

$$\Omega^{0i} = \Omega^{ik}\xi^k, \qquad \Omega^{00} = \Omega^{ik}\xi^i\xi^k. \tag{45.11}$$

Introducing (45.6) in (44.9) and retaining only the terms of order $1/r$ in $v^{\lambda\mu}{}_{,\nu}$ we see at once that, because of (45.9) and (45.10), the expression (44.9) for $\mathrm{t}_\lambda'^{\mu}$ reduces to its first term:

$$\mathrm{t}_\lambda'^{\mu} \to t_\lambda{}^\mu = \frac{1}{4\kappa}\,\eta^{\mu\alpha}\left(\eta_{\varrho\sigma}\eta_{\nu\beta} - \tfrac{1}{2}\eta_{\varrho\nu}\eta_{\sigma\beta}\right) v^{\sigma\beta}{}_{,\alpha} v^{\varrho\nu}{}_{,\lambda}\,. \tag{45.12}$$

For the integrand $\mathrm{t}_\lambda'^{i} n^i$ in (44.14) we find, remembering that $n^i=\xi^i$ and $\xi^i\xi^i=1$:

$$\mathrm{t}_\lambda'^{i} n^i = \frac{\kappa}{16\pi^2 r^2}\,\xi_\lambda\, Q; \tag{45.13}$$

$$Q = \left(\eta_{\varrho\alpha}\eta_{\sigma\beta} - \tfrac{1}{2}\eta_{\varrho\sigma}\eta_{\alpha\beta}\right)\Omega^{\varrho\sigma}\Omega^{\alpha\beta}\,. \tag{45.14}$$

The expression (45.13) may be called the 4-dimensional Poynting vector of the gravitational field in the linear approximation. It is, as in the Maxwell theory, parallel to the propagation vector ξ_λ.

We shall now discuss in some detail the quantity Q and we shall prove that

$$Q \geqslant 0\,. \tag{45.15}$$

This result will be obtained if we proceed in the same way as in the case of electromagnetic radiation. We introduce the projection operator relative to the radial direction ξ^i:

$$\zeta^{ik} = \delta^{ik} - \xi^i\xi^k \quad \Rightarrow \quad \zeta^{ik}\xi^k = 0\,, \qquad \zeta^{ik}\zeta^{il} = \zeta^{kl}\,. \tag{45.16}$$

We then define the transverse quantities:

$$\Omega_\perp^{ik} = \zeta^{il}\zeta^{km}\Omega^{lm}\,. \tag{45.17}$$

These quantities have in general a non-vanishing trace:

$$\Omega_\perp^{ii} = \Omega^{ii} - \Omega^{ik}\xi^i\xi^k\,. \tag{45.18}$$

We finally introduce the transverse and traceless quantities:

$$\left.\begin{aligned}\omega^{ik} &= \Omega_\perp^{ik} - \tfrac{1}{2}\zeta^{ik}\Omega_\perp^{mm}\\ &= \zeta^{il}\zeta^{km}\Omega^{lm} - \tfrac{1}{2}\zeta^{ik}\left(\Omega^{ll} - \Omega^{lm}\xi^l\xi^m\right).\end{aligned}\right\} \tag{45.19}$$

Indeed, the quantities ω^{ik} satisfy the relations:

$$\omega^{ik}\xi^k = 0\,, \qquad \omega^{ii} = 0\,. \tag{45.20}$$

Using the expressions (45.19) we verify directly the equation:

$$\omega^{ik}\omega^{ik} \equiv \sum_{i,k}\left(\omega^{ik}\right)^2 = \left(\eta_{\varrho\alpha}\eta_{\sigma\beta} - \tfrac{1}{2}\eta_{\varrho\sigma}\eta_{\alpha\beta}\right)\Omega^{\varrho\sigma}\Omega^{\alpha\beta} = Q\,, \tag{45.21}$$

which proves the relation (45.15). The physical meaning of this relation is that the energy of gravitational radiation is positive, at least in the linear approximation. We shall see in Section 49 that this result follows also from the more exact discussion of gravitational radiation.

Because of the relations (45.20) there are $6-3-1=2$ independent quantities ω^{ik}. Since these quantities give the simplest description of gravitational radiation, we say that this radiation has two degrees of freedom: We have the same number of degrees of freedom for gravitational and for electromagnetic radiation, in spite of the different tensor properties of the two fields.

Returning to Equation (45.5) we write $v^{\lambda\mu}$ in the form:

$$v^{\lambda\mu} = \frac{\kappa}{2\pi r}\,\phi^{\lambda\mu} + O\left(\frac{1}{r^2}\right). \tag{45.22}$$

Since

$$\Omega^{\lambda\mu} = \dot{\phi}^{\lambda\mu}, \tag{45.23}$$

the dot meaning derivative with respect to the retarded time $t-r$, we conclude from (45.10) that:

$$\dot{\phi}^{\lambda\mu}\xi_\mu = 0.$$

Therefore, if we write

$$\phi^{\lambda\mu}\xi_\mu = a^\lambda, \tag{45.24}$$

we shall have:

$$\dot{a}^\lambda = 0.$$

Let us now introduce the quantities

$${}_0F^{\lambda\mu} = \begin{pmatrix} a^0 + a^s\xi^s & a^i \\ a^i & 0 \end{pmatrix}; \qquad F^{\lambda\mu} = \phi^{\lambda\mu} - {}_0F^{\lambda\mu}. \tag{45.25}$$

We then write (45.22) in the form:

$$v^{\lambda\mu} = \frac{\kappa}{2\pi r}\left({}_0F^{\lambda\mu} + F^{\lambda\mu}\right) + O\left(\frac{1}{r^2}\right) \tag{45.26}$$

and we verify immediately the relations:

$${}_0\dot{F}^{\lambda\mu} = 0, \qquad F^{\lambda\mu}\xi_\mu = 0. \tag{45.27}$$

The constant term ${}_0F^{\lambda\mu}$ describes the time-independent background field, while $F^{\lambda\mu}$ describes the emitted gravitational radiation. Note that, because of the second Equation (45.27), the quantities F^{i0} and F^{00} are determined by F^{ik}.

From F^{ik} we can determine quantities f^{ik} in exactly the same way as we derived the quantities ω^{ik} from Ω^{ik}:

$$f^{ik} = \zeta^{il}\zeta^{km}F^{lm} - \tfrac{1}{2}\zeta^{ik}\left(F^{ll} - F^{00}\right). \tag{45.28}$$

The f^{ik} satisfy relations similar to (45.20):

$$f^{ik}\xi^k = 0, \qquad f^{ii} = 0.$$

From (45.28), (45.23) and (45.19) we derive the relation:

$$\dot{f}^{ik} = \omega^{ik}. \tag{45.29}$$

Actually the quantities

$$f^{\lambda\mu} = \begin{pmatrix} 0 & 0 \\ 0 & f^{ik} \end{pmatrix} \tag{45.30}$$

define the simplest, reduced form $\tilde{v}^{\lambda\mu}$ of the time-dependent part of $v^{\lambda\mu}$:

$$\tilde{v}^{\lambda\mu} = \frac{\kappa}{2\pi r} f^{\lambda\mu}. \tag{45.31}$$

One can verify without difficulty that $\tilde{v}^{\lambda\mu}$ can be obtained from the time-dependent part of $v^{\lambda\mu}$ by the coordinate transformation:

$$\tilde{x}^\mu = x^\mu + \varepsilon\xi^\mu, \tag{45.32}$$

the $\varepsilon\xi^\mu$ satisfying the conditions:

$$(\varepsilon\xi^0)_{,0} = \frac{-\kappa}{8\pi r}(F^{00} + F^{ii}), \qquad (\varepsilon\xi^i)_{,0} = (\varepsilon\xi^0)_{,i} - \frac{\kappa}{2\pi r} F^{0i}. \tag{45.33}$$

The quantities $\tilde{v}^{\lambda\mu}$ deserve the name of the *normal form* of the time-dependent part of $v^{\lambda\mu}$ (of order $1/r$) for the following reason. Using (31.8) one can determine the term of order $1/r$ of the Riemann tensor in the linear approximation. The result is given by the following simple relations ($i, k, \ldots = 1, 2, 3$):

$$\left.\begin{aligned} R'_{i0k0} &= \frac{\kappa}{4\pi r}\ddot{f}^{ik} + O\left(\frac{1}{r^2}\right), \\ R'_{ikl0} &= -\frac{\kappa}{4\pi r}(\ddot{f}^{il}\xi^k - \ddot{f}^{kl}\xi^i) + O\left(\frac{1}{r^2}\right), \\ R'_{iklm} &= -\frac{\kappa}{4\pi r}\varepsilon_{ikn}\varepsilon_{lms}\ddot{f}^{ns} + O\left(\frac{1}{r^2}\right); \end{aligned}\right\} \tag{45.34}$$

ε_{ikl} is the 3-dimensional permutation symbol defined by (32.30).

We thus see that the quantities f^{ik}, which determine the gravitational Poynting vector according to (45.21) and (45.29), determine also the Riemann tensor in a straightforward manner, as shown by (45.34). One may therefore be tempted to consider the Poynting vector (45.13) as meaningful not only for calculating the total energy and momentum of the radiation, but also for determining its distribution in space-time.

46. Gravitational Multipoles. The 4-Pole Radiation

In this section we shall examine briefly the decomposition of gravitational radiation in multipoles, following the similar method developed for the study of electromagnetic radiation. We shall then derive the formulae for the energy and momentum of the simplest type of gravitational radiation, which is the 4-pole radiation.

From (45.3) we derive, for $r \gg |X^i|$, the relation:

$$R = r\left(1 - 2\frac{x^i X^i}{r^2} + \frac{X^i X^i}{r^2}\right)^{1/2} = r - \xi^i X^i + O\left(\frac{1}{r}\right). \tag{46.1}$$

Introducing this value of R into (45.2) we find:

$$T = T_0 + \xi^i X^i + O\left(\frac{1}{r}\right); \qquad T_0 = t - r. \tag{46.2}$$

This last relation allows us to develop the quantity ${}_0\mathfrak{T}^{\lambda\mu}(X^i, T)$, considered as a function of the variable T, into a Taylor series around the value $T = T_0$ of the variable:

$$\left.\begin{aligned} {}_0\mathfrak{T}^{\lambda\mu}(T) &= {}_0\mathfrak{T}^{\lambda\mu}(T_0) + \frac{\partial}{\partial T_0}\, {}_0\mathfrak{T}^{\lambda\mu}(T_0)\cdot \xi^i X^i + \\ &+ \frac{1}{2}\frac{\partial^2}{\partial T_0^2}\, {}_0\mathfrak{T}^{\lambda\mu}(T_0)\cdot \xi^i \xi^k X^i X^k + \cdots. \end{aligned}\right\} \tag{46.3}$$

Introducing the expansion (46.3) into the integral appearing in (45.5) we find:

$$\left.\begin{aligned} \int_C {}_0\mathfrak{T}^{ik}(T)\, \mathrm{d}^3X &= \int_\Sigma {}_0\mathfrak{T}^{ik}(T_0)\, \mathrm{d}^3X + \\ &+ \xi^l \frac{\mathrm{d}}{\mathrm{d}T_0}\int_\Sigma {}_0\mathfrak{T}^{ik}(T_0)\, X^l\, \mathrm{d}^3X + \\ &+ \tfrac{1}{2}\xi^l \xi^m \frac{\mathrm{d}^2}{\mathrm{d}T_0^2}\int_\Sigma {}_0\mathfrak{T}^{ik}(T_0)\, X^l X^m\, \mathrm{d}^3X + \cdots. \end{aligned}\right\} \tag{46.4}$$

According to the definition (45.2) of T the integral on the left-hand side of (46.4) is an integral over the past light cone C of the point (x^i, t). On the contrary, in the right-hand side of (46.4) we have integrals over the space-like hypersurface Σ defined by the equation $t - r = \text{const}$. Equation (46.4) is the basic relation for the multipole expansion of gravitational radiation.

We have to stress that the expansion (46.3) will converge quickly and consequently the multipole expansion (46.4) will be really useful if ${}_0\mathfrak{T}^{\lambda\mu}$ is a quasi-periodic function of T, with the average period $\mathfrak{t}$ satisfying the condition:

$$|\xi^i X^i| \ll c\mathfrak{t}. \tag{46.5}$$

The factor ξ^i is of the order 1 and consequently $\xi^i X^i$ is of the order of $|X^i|$, i.e. of the same order as the linear dimension a of the source of the radiation appearing in (44.10). Remembering that ct is the (average) wavelength λ of the radiation,

$$ct = \lambda, \tag{46.6}$$

we write the condition (46.5) in the more intuitive form:

$$a \ll \lambda. \tag{46.7}$$

We consider the first integral in the right-hand side of Equation (46.4). This integral can be evaluated with the help of the following relation, which is an immediate consequence of the conservation law (44.7):

$${}_0\mathfrak{T}^{\lambda\mu} = \tfrac{1}{2}(X^\lambda X^\mu \, {}_0\mathfrak{T}^{\varrho\sigma})_{,\varrho\sigma}. \tag{46.8}$$

We find:

$$\int_\Sigma {}_0\mathfrak{T}^{ik} \, \mathrm{d}^3 X = \tfrac{1}{2} \int_\Sigma (X^i X^k \, {}_0\mathfrak{T}^{\varrho\sigma})_{,\varrho\sigma} \, \mathrm{d}^3 X .$$

For ϱ or $\sigma \neq 0$ the integral on the right-hand side can be transformed into an integral over the sphere of radius $r \to \infty$ and vanishes because of (44.10). We have finally only the term corresponding to $\varrho = \sigma = 0$. The result is:

$$\int_\Sigma {}_0\mathfrak{T}^{ik} \, \mathrm{d}^3 X = \frac{1}{2} \frac{\mathrm{d}^2}{\mathrm{d}T_0^2} \int_\Sigma {}_0\mathfrak{T}^{00} X^i X^k \, \mathrm{d}^3 X . \tag{46.9}$$

We thus see that the term of lowest order appearing in the multipole expansion of the gravitational radiation is the 4-pole term.

The discussion of the subsequent integrals in the right-hand side of Equation (46.4) shows that these integrals are either 4-pole moments of ${}_0\mathfrak{T}^{0i}$ and of ${}_0\mathfrak{T}^{ik}$ or higher order moments of ${}_0\mathfrak{T}^{00}$. We shall limit ourselves to the 4-pole type of radiation and we shall also assume that the source of the radiation is in a state of non-relativistic motion. In this case we have:

$${}_0\mathfrak{T}^{ik} \ll {}_0\mathfrak{T}^{0i} \ll {}_0\mathfrak{T}^{00},$$

and consequently it is sufficient to consider the integrals (46.9) only. We then find from (45.5):

$$v^{ik} = \frac{\kappa}{4\pi r} \ddot{\mathrm{d}}^{ik} + O\left(\frac{1}{r^2}\right); \tag{46.10}$$

$$\mathrm{d}^{ik} = \int_\Sigma {}_0\mathfrak{T}^{00} X^i X^k \, \mathrm{d}^3 X . \tag{46.11}$$

The value of Ω^{ik} is derived from (45.23):

$$\Omega^{ik} = \tfrac{1}{2} \dddot{\mathrm{d}}^{ik}. \tag{46.12}$$

We now calculate the quantity Q which is given by (45.14). Taking into account the relations (45.11) we find:

$$\left.\begin{aligned} Q = \Omega^{ik}\Omega^{ik} - 2\xi^i\xi^l\Omega^{kl}\Omega^{ki} + \tfrac{1}{2}\xi^i\xi^k\xi^l\xi^m\Omega^{ik}\Omega^{lm} \\ - \tfrac{1}{2}\Omega^{ii}\Omega^{kk} + \xi^k\xi^l\Omega^{ii}\Omega^{kl}. \end{aligned}\right\} \quad (46.13)$$

The energy of the radiation is determined by Equation (44.14) with $\mu=0$. Remembering that Ω^{ik} is, according to (46.12), a function of the retarded time $T_0=t-r$ only, we see at once that for the calculation of the energy of the radiation we have to integrate over the sphere $r=\text{const}$ expressions of the forms $\xi^i\xi^k$ and $\xi^i\xi^k\xi^l\xi^m$. The values of these integrals are obtained easily in polar coordinates:

$$\xi^1 = \sin\theta\cos\varphi, \quad \xi^2 = \sin\theta\sin\varphi, \quad \xi^3 = \cos\theta; \quad \mathrm{d}S = r^2\sin\theta\,\mathrm{d}\theta\,\mathrm{d}\varphi.$$

An elementary calculation shows then that the average values of these expressions over the sphere $r=\text{const}$ are:

$$\overline{\xi^i\xi^k} = \tfrac{1}{3}\delta^{ik}, \qquad \overline{\xi^i\xi^k\xi^l\xi^m} = \tfrac{1}{15}(\delta^{ik}\delta^{lm} + \delta^{il}\delta^{km} + \delta^{im}\delta^{kl}). \quad (46.14)$$

Introducing these expressions in (46.13) we find:

$$\bar{Q} = \tfrac{2}{5}(\Omega^{ik} - \tfrac{1}{3}\delta^{ik}\Omega^{ll})(\Omega^{ik} - \tfrac{1}{3}\delta^{ik}\Omega^{mm}) = \tfrac{1}{10}\dddot{D}^{ik}\dddot{D}^{ik}. \quad (46.15)$$

The quantities D^{ik} are the traceless 4-pole moments of the material energy-density ${}_0\mathfrak{T}^{00}$:

$$D^{ik} = \mathrm{d}^{ik} - \tfrac{1}{3}\delta^{ik}\,\mathrm{d}^{ll}. \quad (46.16)$$

With this value of $\bar{Q}$ we find for the energy of the radiation which crosses the sphere S of radius $r=\text{const}\to\infty$ per unit time:

$$\frac{\mathrm{d}R_0}{\mathrm{d}t} \equiv \frac{\mathrm{d}E}{\mathrm{d}t} = \frac{G}{5c^5}\dddot{D}^{ik}\dddot{D}^{ik}. \quad (46.17)$$

We have reintroduced in this formula the missing factors c. Equation (46.17) was obtained by Einstein in 1916.

For the momentum of the radiation we have to use (44.14) with $\mu=i$. We then see that each term of the integrand contains an odd number of factors ξ^i and consequently all integrals over the sphere S vanish: Because of its symmetry the 4-pole gravitational radiation has no total momentum,

$$\frac{\mathrm{d}R_i}{\mathrm{d}t} = 0. \quad (46.18)$$

47. Gravitational Shock Waves. General Discussion

We start with an intuitive example taken from the Maxwell theory. Let us consider a point charge which is at rest for $t<0$ and starts being accelerated at $t=0$ (Figure 30).

The retarded field of this particle is stationary initially, but variable later. The hypersurface Σ which separates the regions of the stationary and the variable field is seen at once to be the future light cone of the point O (Figure 30). The surface Σ is the *front* of the electromagnetic wave which is emitted by the accelerated particle.

Mathematically the surface Σ is characterized by the fact that certain derivatives of the electromagnetic potential A_μ are discontinuous across Σ. This follows at once from the remark that the derivatives of A_μ with respect to the time coordinate x^0 are zero below Σ, but some of these derivatives will certainly be non-zero above Σ. These discontinuities are propagated along Σ in a way determined by the Maxwell equations. We say that there is an electromagnetic *shock wave* propagated on Σ.

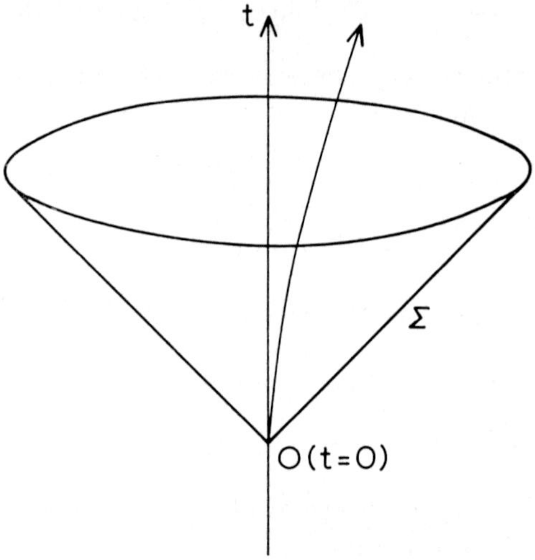

Fig. 30. The simplest type of an electromagnetic shock wave.

Shock waves exist in any field theory having hyperbolic field equations and consequently also in general relativity. We shall discuss in detail *a gravitational shock wave of order* $n=2$, which is defined in the following way. Let the hypersurface Σ be defined by the equation:

$$z(x^\alpha) = 0. \tag{47.1}$$

On either side of Σ we have the half-spaces M^- and M^+ (Figure 31). Consider a pair of points P^- and P^+, both tending to a point P on Σ. We introduce the following assumptions:

$$[g_{\lambda\mu}] \equiv \lim_{P^- \text{ and } P^+ \to P} \{g_{\lambda\mu}(P^+) - g_{\lambda\mu}(P^-)\} = 0; \tag{47.2}$$

$$[g_{\lambda\mu,\nu}] = 0\,; \tag{47.3}$$

$$[g_{\lambda\mu,\nu\varrho}] \neq 0 \quad \text{for at least some} \quad g_{\lambda\mu,\nu\varrho}. \tag{47.4}$$

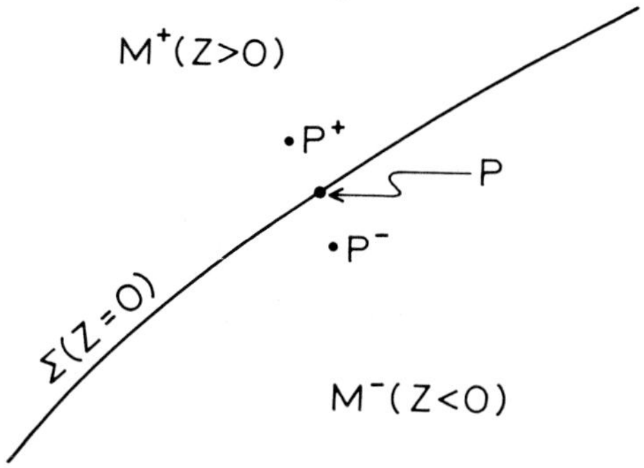

Fig. 31. Defining discontinuities across a hypersurface Σ.

Further we shall assume that the hypersurface Σ lies in a region of space in which the vacuum field equations,

$$R_{\lambda\mu} = 0, \tag{47.5}$$

are valid. We shall determine the properties of the gravitational shock wave which follow from Equation (47.5).

In order to simplify some calculations we introduce auxiliary variables $\tilde{x}^\mu$ defined by the equations:

$$\tilde{x}^0 = z(x^\alpha), \qquad \tilde{x}^i = f^i(x^\alpha). \tag{47.6}$$

The equation of Σ can now be written in the form:

$$\tilde{x}^0 = 0. \tag{47.7}$$

We shall assume that the functions $z(x^\alpha)$ and $f^i(x^\alpha)$ have first and second derivatives which are continuous across Σ. Moreover we have to demand:

$$\det \frac{\partial \tilde{x}}{\partial x} \neq 0.$$

If a function $F(\tilde{x}^\alpha)$ is continuous on Σ,

$$[F] = 0,$$

and provided that the derivatives

$$F_{,\tilde{i}} \equiv \frac{\partial F}{\partial \tilde{x}^i}, \qquad F_{,\tilde{i}\tilde{k}} \quad \text{etc.}$$

exist, one can see immediately that these derivatives are continuous on Σ:

$$[F_{,\tilde{i}}] = 0, \qquad [F_{,\tilde{i}\tilde{k}}] = 0 \quad \text{etc.} \tag{47.8}$$

From (47.2) and (47.8) we get:

$$[g_{\lambda\mu,\tilde{i}}] = 0 = [g_{\lambda\mu,\tilde{i}\tilde{k}}] = \cdots .$$

Similarly from (47.3) and (47.8):

$$[g_{\lambda\mu,\tilde{o}}]=0\,;\qquad [g_{\lambda\mu,\tilde{o}\tilde{i}}]=0=[g_{\lambda\mu,\tilde{o}\tilde{i}\tilde{k}}]=\cdots.$$

But according to (47.4) we shall have:

$$[g_{\lambda\mu,\tilde{o}\tilde{o}}]=\gamma_{\lambda\mu}\quad\text{with at least some}\quad \gamma_{\lambda\mu}\neq 0\,. \tag{47.9}$$

Thus derivatives with at most one index equal to $\tilde{o}$ are continuous on Σ (if they exist); but derivatives with two or more indices equal to $\tilde{o}$ will be discontinuous.

Returning to the initial variables x^λ we write the elementary formula:

$$g_{\lambda\mu,\nu\varrho}=g_{\lambda\mu,\tilde{\alpha}\tilde{\beta}}\frac{\partial\tilde{x}^\alpha}{\partial x^\nu}\frac{\partial\tilde{x}^\beta}{\partial x^\varrho}+g_{\lambda\mu,\tilde{\alpha}}\frac{\partial^2\tilde{x}^\alpha}{\partial x^\nu\partial x^\varrho}\,.$$

It follows then from (47.9):

$$[g_{\lambda\mu,\nu\varrho}]=\gamma_{\lambda\mu}p_\nu p_\varrho\,;\qquad p_\nu=\frac{\partial\tilde{x}^0}{\partial x^\nu}\,. \tag{47.10}$$

According to the first of Equations (47.6) we have:

$$p_\nu=\frac{\partial z}{\partial x^\nu}\,, \tag{47.11}$$

i.e. p_ν is the normal to the surface Σ.

The vacuum field equations (47.5) can be written, using (14.2), in the form:

$$g^{\alpha\beta}(g_{\alpha\beta,\lambda\mu}+g_{\lambda\mu,\alpha\beta}-g_{\alpha\lambda,\mu\beta}-g_{\mu\beta,\alpha\lambda})+\cdots=0\,. \tag{47.12}$$

The omitted terms depend only on $g_{\nu\varrho}$ and $g_{\nu\varrho,\sigma}$. Since Equation (47.12) is valid on both sides of Σ:

$$R_{\lambda\mu}(P^-)=0=R_{\lambda\mu}(P^+)\,,$$

we shall have:

$$[R_{\lambda\mu}]=0\,. \tag{47.13}$$

Since the $g_{\nu\varrho}$ and $g_{\nu\varrho,\sigma}$ are continuous on Σ, we find from (47.10) and (47.13):

$$g^{\alpha\beta}(\gamma_{\alpha\beta}p_\lambda p_\mu+\gamma_{\lambda\mu}p_\alpha p_\beta-\gamma_{\alpha\lambda}p_\mu p_\beta-\gamma_{\alpha\mu}p_\lambda p_\beta)=0\,. \tag{47.14}$$

This is the condition derived from the vacuum field equation (47.5) for the discontinuities $\gamma_{\lambda\mu}$. We shall discuss this equation in detail.

The conclusions drawn from (47.14) are basically different in the following two possible cases:

$$g^{\alpha\beta}p_\alpha p_\beta\equiv p^\alpha p_\alpha\neq 0\quad\text{or}\quad p^\alpha p_\alpha=0\,. \tag{47.15}$$

We have therefore to consider each one of these cases separately.

We start with the first of the cases (47.15). Equation (47.14) can then be solved with respect to $\gamma_{\lambda\mu}$. The result is:

$$\gamma_{\lambda\mu} = p_\lambda a_\mu + p_\mu a_\lambda , \tag{47.16}$$

the vector a_λ being given by the relation:

$$a_\lambda = (\gamma_{\alpha\lambda} p^\alpha - \tfrac{1}{2} g^{\alpha\beta} \gamma_{\alpha\beta} p_\lambda) \cdot (p^\mu p_\mu)^{-1} .$$

One can prove easily that discontinuities of the form (47.16) are trivial in the following sense: They can be eliminated or produced by coordinate transformations

$$x^{*\alpha} = \varphi^\alpha (x^\beta) \tag{47.16$'$}$$

in which the functions $\varphi^\alpha (x^\beta)$ have discontinuous third derivatives. This conclusion is in agreement with the following remark: If we calculate the discontinuities of the components of the Riemann tensor, using Equations (14.2) and (47.16), we find that they vanish:

$$[R_{\lambda\mu\nu\varrho}] = 0 . \tag{47.17}$$

Geometrically the first of Equations (47.15) means that the vector p_ν is time- or space-like and consequently the surface Σ is space- or time-like. The preceding result means that in this case the field equations do not allow essential (i.e. non-trivial) discontinuities of the field variables. Indeed, this is the case in which we can formulate and solve the *Cauchy problem* of general relativity: If we are given on Σ the values of $g_{\lambda\mu}$ and $g_{\lambda\mu,\nu}$, we can determine from the field equations the second and higher derivatives of $g_{\lambda\mu}$ and in this way we can determine the field in the neighbourhood of Σ.

We now turn to the second case (47.15), in which p_ν is a null vector. We shall assume that this is true at all points of Σ. The hypersurface Σ is then called a *characteristic* or *null surface*. Equation (47.14) is now simplified to the form:

$$\gamma^\alpha{}_\alpha p_\lambda p_\mu - \gamma_{\alpha\lambda} p^\alpha p_\mu - \gamma_{\alpha\mu} p^\alpha p_\lambda = 0 . \tag{47.18}$$

Introducing the vector

$$b_\lambda = \gamma_{\lambda\alpha} p^\alpha - \tfrac{1}{2} \gamma^\alpha{}_\alpha p_\lambda \tag{47.19}$$

we write Equation (47.18) in the form:

$$b_\lambda p_\mu + b_\mu p_\lambda = 0 . \tag{47.20}$$

It follows that Equation (47.18) is equivalent to the following 4 equations:

$$b_\lambda = 0 \quad \Leftrightarrow \quad \gamma_{\lambda\alpha} p^\alpha = \tfrac{1}{2} \gamma^\alpha{}_\alpha p_\lambda . \tag{47.21}$$

There are 10 quantities $\gamma_{\lambda\mu}$ and only the 4 conditions (47.21) to be satisfied by them. Therefore the $\gamma_{\lambda\mu}$ will contain in this case 6 arbitrary quantities. Actually 4 of these arbitrary quantities are the a_μ appearing in (47.16). Indeed, one can verify directly that the expression (47.16) satisfies Equation (47.21). Therefore we can write the general solution of Equation (47.21) in the form

$$\gamma_{\lambda\mu} = a_\lambda p_\mu + a_\mu p_\lambda + \gamma'_{\lambda\mu}, \tag{47.22}$$

the $\gamma'_{\lambda\mu}$ containing now only two arbitrary quantities. From the point of view of the shock wave the part $a_\lambda p_\mu + a_\mu p_\lambda$ of $\gamma_{\lambda\mu}$ in (47.22) is uninteresting, because it can be eliminated by a transformation of the type (47.16′) with discontinuous third derivatives. It is the part $\gamma'_{\lambda\mu}$ which describes the *essential* discontinuities.

The detailed structure of $\gamma'_{\lambda\mu}$ will be determined in the next section. Here we shall discuss in some detail the geometrical meaning of the second Equation (47.15).

The vector p_λ is the normal to the surface Σ. Consequently p_λ is orthogonal to any vector A^λ lying on Σ and inversely any vector which is orthogonal to p_λ lies on Σ. It follows that p_λ itself lies on Σ, this being possible because of the indefinite character of the metric $g_{\lambda\mu}$. Surfaces Σ with this property have to be excluded in the discussion of the Cauchy problem of general relativity and are called *characteristic surfaces* of the field equations.

The vector p_λ determines a vector-field on Σ. The trajectories of this vector field are evidently null-lines. The following theorem can be proved easily: *The trajectories of the vector field p_λ on Σ are null geodesics of the 4-dimensional space-time.* In other words a null geodesic passing through a point P of Σ and having there the direction of the vector p_λ lies entirely on Σ. These null-geodesics are called the *bicharacteristics* of Σ.

At a point P of Σ there is only one null direction lying on Σ, which is the direction of the vector p_λ corresponding to P. It follows that the light cone of P will be tangent to Σ. One can see easily that the future light cone of P will be entirely on the one side of Σ and the past light cone on the other side.

A characteristic surface Σ is degenerate from the following point of view. A non-characteristic surface allows to determine at any one of its points 4 independent directions: the normal to the surface and 3 independent directions on the surface. Consequently such a surface allows to define a complete frame or a *tetrad* of the space-time. On the contrary, when we have a characteristic surface Σ, its normal p_λ lies on Σ. Consequently there are besides p_λ only two other independent directions, for example the vectors a^λ and b^λ satisfying the relations:

$$a^\lambda p_\lambda = b^\lambda p_\lambda = 0. \tag{47.23}$$

Therefore a characteristic surface Σ cannot determine a complete tetrad of the space-time, but only a triad on Σ. One can prove easily that vectors a^λ and b^λ, which are orthogonal (but not collinear) to a null vector p_λ, are necessarily space-like.

48. Local Structure and Propagation of Discontinuities

In order to determine the form of the quantity $\gamma'_{\lambda\mu}$ appearing in (47.22) it is necessary to define a triad on Σ by choosing at each point of Σ two space-like vectors a^λ and

b^{λ}. It will be convenient to take a^{λ} and b^{λ} orthonormal:

$$a^{\lambda} b_{\lambda} = 0, \qquad a^{\lambda} a_{\lambda} = b^{\lambda} b_{\lambda} = -1. \tag{48.1}$$

Of course, a^{λ} and b^{λ} will satisfy (47.23) as they are on Σ.

It is possible to choose vectors a^{λ} and b^{λ}, which satisfy the conditions (48.1) but are otherwise arbitrary on a 2-dimensional surface cutting all bicharacteristics of Σ, and then to define these vectors on the entire surface Σ by parallel transport along the bicharacteristics. We shall then have the relations:

$$a^{\lambda}{}_{;\mu} p^{\mu} = 0 = b^{\lambda}{}_{;\mu} p^{\mu}, \tag{48.2}$$

which are useful because they simplify some of the subsequent formulae.

The detailed discussion of the quantity $\gamma'_{\mu\nu}$ has shown that it can be expressed with the help of two vectors A_{μ} and B_{μ}:

$$\gamma'_{\mu\nu} = A_{\mu} B_{\nu} + A_{\nu} B_{\mu}. \tag{48.3}$$

The vectors A_{μ} and B_{μ} are both on the 2-dimensional surface spanned by a_{μ} and b_{μ}:

$$A_{\mu} = a' a_{\mu} + b' b_{\mu}, \qquad B_{\mu} = a'' a_{\mu} + b'' b_{\mu}, \tag{48.4}$$

and they can be chosen so as to be mutually orthogonal and of equal length:

$$A_{\lambda} B^{\lambda} = 0, \qquad A_{\lambda} A^{\lambda} = B_{\lambda} B^{\lambda}. \tag{48.5}$$

According to (48.4) the vectors A_{λ} and B_{λ} contain the 4 coefficients $a', \ldots b''$. Since we have to satisfy the 2 conditions (48.5), we see that $\gamma'_{\lambda\mu}$ contains 2 arbitrary quantities.

We obtain a more convenient representation of the vectors A_{λ} and B_{λ} if we introduce the complex null vector

$$m^{\lambda} = \frac{1}{\sqrt{2}} (a^{\lambda} + i b^{\lambda}). \tag{48.6}$$

Indeed, one verifies directly the relation:

$$m^{\lambda} m_{\lambda} = 0. \tag{48.7}$$

Moreover one finds:

$$m^{\lambda} \bar{m}_{\lambda} = -1. \tag{48.8}$$

We now write the vectors A_{λ} and B_{λ} in the form:

$$A_{\lambda} = P (e^{i\delta} m_{\lambda} + e^{-i\delta} \bar{m}_{\lambda}), \qquad B_{\lambda} = \frac{P}{i} (e^{i\delta} m_{\lambda} - e^{-i\delta} \bar{m}_{\lambda}), \tag{48.9}$$

containing the two real quantities P and δ, which are respectively the amplitude factor and the phase angle of the shock wave.

Omitting the trivial part $a_{\lambda} p_{\mu} + a_{\mu} p_{\lambda}$ in (47.22) we write:

$$\gamma_{\lambda\mu} = \gamma'_{\lambda\mu} = A_{\lambda} B_{\mu} + A_{\mu} B_{\lambda}. \tag{48.10}$$

We recapitulate the properties of the so reduced $\gamma_{\lambda\mu}$. It is orthogonal to p_λ and traceless:

$$\gamma_{\lambda\mu}p^\mu = 0 = g^{\lambda\mu}\gamma_{\lambda\mu}. \tag{48.11}$$

We conclude that gravitational shock waves are *transverse* and have 2 degrees of freedom. It is to be noted that these qualitative properties are identical with those of electromagnetic shock waves. We may add that the transversality of the shock waves and more generally of the radiation is the deeper reason for the validity of the Birkhoff theorem established in Section 21 for the gravitational field and in Section 37 for the combined gravitational and electromagnetic field.

The results obtained until now have been derived from local considerations, at one point P of Σ, and therefore they describe the *local structure* of the shock wave. We get important additional information on how the discontinuities are propagated on Σ if we consider the field equations (47.5) derived with respect to x^ν:

$$R_{\lambda\mu,\nu} = 0. \tag{48.12}$$

By requiring this equation to be satisfied on both sides of Σ we are led to the condition

$$[R_{\lambda\mu,\nu}] = 0. \tag{48.13}$$

The detailed discussion of (48.13) leads to *propagation relations* for the discontinuities along the bicharacteristics of Σ.

The calculations leading to the propagation relations are rather long and we shall not reproduce them here. We shall only discuss briefly a supplementary condition which simplifies the final results and will also be needed in the next section for another purpose. We shall arrive at this condition if we remark that the equation

$$z(x^\alpha) = \text{const} \equiv k$$

describes, for different values of k, a family of hypersurfaces which contains Σ. But these hypersurfaces are in general not null surfaces when $k \neq 0$:

$$p_\lambda p^\lambda \neq 0 \quad \text{for} \quad k \neq 0.$$

It follows that in general we shall have

$$(p_\lambda p^\lambda)_{,\mu} \neq 0$$

even on the null-surface Σ corresponding to $k = 0$.

It can now be shown easily that by rewriting the equation of Σ in the form:

$$\tilde{z}(x^\alpha) = 0; \qquad \tilde{z}(x^\alpha) = z(x^\alpha)f(x^\alpha), \tag{48.14}$$

where $f(x^\alpha)$ is a function which does not vanish on Σ, one can obtain by an appropriate choice of $f(x^\alpha)$ that the new null vector

$$\tilde{p}_\lambda = \frac{\partial \tilde{z}}{\partial x^\lambda} \tag{48.15}$$

satisfies the relation:

$$(\tilde{p}_\lambda \tilde{p}^\lambda)_{,\mu} = 0 \quad \text{on} \quad \Sigma. \tag{48.16}$$

We shall assume that the condition (48.16) has been satisfied and we shall write again p_λ instead of $\tilde{p}_\lambda$. The propagation relations obtained from Equation (48.13) take then the following simple form:

$$(P^4 \sqrt{-g}\, p^\lambda)_{,\lambda} = 0\,; \qquad \delta_{,\lambda} p^\lambda = 0\,. \tag{48.17}$$

The first of these equations has the form of the continuity equation and expresses the conservation of the flux of the vector $P^4 p^\lambda$ through the sections of a tube formed by bicharacteristics. It follows that, if $P \neq 0$ at some point Q of a bicharacteristic, we shall have $P \neq 0$ at any other point of the bicharacteristic which is still in the vacuum. Thus a shock wave can only start or end in regions containing sources of the field.

The second Equation (48.17) states that the phase δ remains constant along each bicharacteristic.

We end this section with some general remarks. We have discussed here in some detail a shock wave of order 2. There can be also shock waves of order $n > 2$, in which the discontinuous derivatives are, by definition, of order $\geqslant n$. This case is treated in a similar way as the case $n = 2$, the only difference being that the discussion is now based on the Equations (47.5) and (48.13) derived $n - 2$ times with respect to the coordinates x^λ. The final results of the discussion are identical with those found here for the case $n = 2$. The only difference is that the discontinuity tensor $\gamma_{\lambda\mu}$ determines now the discontinuities of the derivatives of order n:

$$[g_{\lambda\mu,\nu\varrho\ldots}] = \gamma_{\lambda\mu} p_\nu p_\varrho \ldots . \tag{48.18}$$

Shock waves of order $n = 1$ are also possible. Their discussion is different from the one we gave here, but the final results are again found to be of exactly the same form as in the case $n = 2$.

Finally we stress the fact that gravitational shock waves have been discussed on the basis of the exact field equations. This has been possible because the field equations are linear in the second derivatives of $g_{\lambda\mu}$. Since nearly all other questions related to gravitational radiation have to be treated by approximation methods, the exact results obtained from the discussion of gravitational shock waves are particularly valuable.

49. Radiation Coordinates (Bondi Coordinates)

The discussion of Section 47 shows that the description of a shock wave will be simpler if we use coordinates such that the null surface Σ has an equation of the form

$$x^0 = 0\,. \tag{49.1}$$

Indeed, in this case the parametric lines of the coordinates x^i ($i = 1, 2, 3$) lie on Σ and consequently discontinuities will appear only in the derivatives with respect to x^0. Moreover we found in Section 48 that, in order to obtain the simple form (48.17) of the propagation relations, we had to impose the condition (48.16):

$$(p_\lambda p^\lambda)_{,\mu} = 0. \tag{49.2}$$

In the coordinate system in which Σ has the Equation (49.1) the components $\mu = i$ of the condition (49.2) are satisfied automatically. The component $\mu = 0$ is seen easily to have the following qualitative meaning: In the family of hypersurfaces determined by the equation

$$x^0 = \text{const} \tag{49.3}$$

the next neighbours to Σ are also null surfaces.

These remarks suggest us to construct coordinates in which *all* the hypersurfaces determined by (49.3) are null-surfaces. One is led in this way to the *Bondi* or *radiation coordinates*, which have been found to be especially suited to the discussion of the propagation of gravitational radiation. In this section we shall describe briefly the main features of these coordinates.

The construction of the radiation coordinates is based on a family of non-intersecting null surfaces. The first coordinate x^0 is defined by the requirement that this family of surfaces have the Equation (49.3). The surface Σ on which is propagated the shock wave representing the front of the radiation has of course to be included in this family.

We shall consider gravitational fields which have sources of finite extension in the 3-dimensional space and satisfy the condition (20.11) of limiting Minkowski metric at $r \to \infty$. A shock wave starting from such sources propagates on a null surface Σ which tends asymptotically, for $r \to \infty$, to a future light cone of the limiting Minkowski metric. We shall demand that all the null surfaces determined by (49.3) have this property.

It has been shown in Section 47 that on any null surface there is a congruence of null geodesics – the bicharacteristics of the null surface. This gives us the possibility to define the second coordinate x^1 again in a geometrical way: We take the bicharacteristics of the null surfaces determined by (49.3) as the parametric lines of the coordinate x^1; i.e. on each bicharacteristic we shall have not only $x^0 = \text{const}$, but also x^2 and $x^3 = \text{const}$. This does not define sufficiently the coordinate x^1, as there is an infinity of possible parameters on a given curve. We shall use here the definition proposed by Newman and Penrose: x^1 is the affine parameter of the bicharacteristics which tends asymptotically to the Minkowskian radial distance at $r \to \infty$.

The remaining two coordinates are generalized polar angles, like the angles θ and φ on a sphere, which have to label in some way the bicharacteristics of the null surface $x^0 = \text{const}$. The simplest possibility is to define the coordinates x^2 and x^3 so that they coincide asymptotically, at $r \to \infty$, with the usual polar angles θ and φ on the sphere $x^1 \equiv r = \text{const}$.

We shall determine the general form of the metric $g_{\lambda\mu}$ corresponding to the so defined coordinates. The normal to the null surface $x^0 = \text{const}$ is the vector:

$$l_\mu = (1, 0, 0, 0). \tag{49.4}$$

The contravariant components of this vector are:

$$l^\mu = g^{\mu\nu} l_\nu = g^{\mu 0}. \tag{49.5}$$

The vector l^μ is tangent to the bicharacteristics, which are parametric lines of the coordinate x^1. Therefore:

$$l^0 = l^2 = l^3 = 0 \quad \Leftrightarrow \quad g^{00} = g^{02} = g^{03} = 0. \tag{49.6}$$

Moreover we shall have:

$$l^1 = \frac{\mathrm{d}x^1}{\mathrm{d}\lambda},$$

λ being the affine parameter on the bicharacteristic. Since we decided to take $\lambda = x^1$, we shall have:

$$l^1 = 1 \quad \Leftrightarrow \quad g^{01} = 1. \tag{49.7}$$

The relations (49.6) and (49.7) are characteristic of the metric in radiation coordinates.

In matrix form we have:

$$g^{\lambda\mu} = \begin{pmatrix} 0 & 1 & 0 & 0 \\ 1 & g^{11} & g^{12} & g^{13} \\ 0 & g^{12} & g^{22} & g^{23} \\ 0 & g^{13} & g^{23} & g^{33} \end{pmatrix}. \tag{49.8}$$

It is easy to calculate from (49.8) and (11.8) the components $g_{\lambda\mu}$. The relations corresponding to (49.6) and (49.7) are now:

$$g_{01} = 1, \qquad g_{11} = g_{12} = g_{13} = 0. \tag{49.9}$$

As an example we determine the radiation form of the Schwarzschild metric. We start from the Eddington form $(26.3)_+$:

$$\mathrm{d}s^2 = \left(1 - \frac{2m}{r}\right)\mathrm{d}t^2 - \left(1 + \frac{2m}{r}\right)\mathrm{d}r^2 + \frac{4m}{r}\,\mathrm{d}t\,\mathrm{d}r - r^2(\mathrm{d}\theta^2 + \sin^2\theta\,\mathrm{d}\varphi^2). \tag{49.10}$$

According to the results obtained in Section 26 the surfaces defined by the equation

$$t - r = \text{const} \tag{49.11}$$

are null-surfaces and r is an affine parameter of the radial null-geodesics defined by (49.11). We therefore can put:

$$t - r = x^0 \equiv u, \qquad r = x^1. \tag{49.12}$$

Introducing (49.12) into (49.10) we find the radiative form of the Schwarzschild metric:

$$\mathrm{d}s^2 = \left(1 - \frac{2m}{r}\right)\mathrm{d}u^2 + 2\,\mathrm{d}u\,\mathrm{d}r - r^2(\mathrm{d}\theta^2 + \sin^2\theta\,\mathrm{d}\varphi^2). \tag{49.13}$$

Putting in (49.13) $m=0$ we get the radiation form of the Minkowski metric ${}_{\mathrm{M}}g_{\lambda\mu}$. We have:

$$ {}_{\mathrm{M}}\mathrm{d}s^2 = \mathrm{d}u^2 + 2\,\mathrm{d}u\,\mathrm{d}r - r^2(\mathrm{d}\theta^2 + \sin^2\theta\,\mathrm{d}\varphi^2). \tag{49.13'} $$

We then find from the requirement:

$$ g_{\lambda\mu} \to {}_{\mathrm{M}}g_{\lambda\mu} \quad \text{for} \quad r \to \infty $$

the following asymptotic relations:

$$ g_{00} = 1 + O\left(\frac{1}{r}\right), \qquad g_{22} = -r^2 + O(r), \qquad g_{33} = -r^2\sin^2\theta + O(r). \tag{49.14} $$

The asymptotic behaviour of the remaining 3 components $g_{\lambda\mu}$ has to be determined by a more detailed analysis, of which we give only the final result:

$$ g_{02} = O(1), \qquad g_{03} = O(1), \qquad g_{23} = O(r). \tag{49.15} $$

We mention, without going into any details, that the coordinate transformations, which preserve the radiative form of the metric – characterized by the relations (49.9), (49.14) and (49.15) – form the *Bondi-Metzner-Sachs group*.

The use of the radiation coordinates in the discussion of the propagation of gravitational radiation offers the following advantages:

(i) It is not necessary to assume that the field is weak everywhere.

(ii) No linearization of the field equations is needed.

(iii) The discussion is based directly on the field equations, with no use of any pseudotensorial relations.

There is, however, an important limitation: It has to be expected that in the neighbourhood of the sources of the field the family of null-surfaces determined by Equation (49.3) will in general develop caustic surfaces. It follows that the construction of radiation coordinates will be possible at some distance from the sources. Actually the method of radiation coordinates has been used only for studying the propagation of gravitational radiation at large distance from its sources.

Bondi applied this method to the case of an axially symmetric field having no angular momentum. We shall not enter into a detailed description of the method. We mention only that it was possible to derive a formula for the energy E of the gravitational radiation. The energy emitted per unit time is given by an integral of a positive definite quantity over the 2-dimensional sphere of radius $r \to \infty$:

$$ \frac{\mathrm{d}E}{\mathrm{d}t} = \frac{1}{4\pi}\int ({}_1\gamma_{,0})^2\,\mathrm{d}\omega; \tag{49.16} $$

$\mathrm{d}\omega$ is the element of solid angle and ${}_1\gamma$ the coefficient of the term of order $1/r$ of a function γ,

$$ \gamma = \frac{1}{r}\,{}_1\gamma(u, \theta) + O\left(\frac{1}{r^2}\right), \tag{49.17} $$

which determines g_{22} according to the formula:

$$g_{22} = - r^2 e^{2\gamma} . \tag{49.18}$$

The relation (49.16) is of the same type as the component $\mu = 0$ of Equation (44.15), because of (45.13) and (45.21). The relation (49.16) has, however, the advantage of being derived directly from the field equations, without linearization. We may add that here also the result (49.16) was obtained with the help of the assumption that the field is stationary initially and finally. The assumption of an initially stationary field is indispensable, because only in this case we can be sure that the field does not contain any incoming radiation.

The detailed discussion of the field equations in radiation coordinates has shown that the function ${}_1\gamma$ is the only field variable on which the field equations do not impose any restriction. Consequently ${}_1\gamma$ is an arbitrary function of u and θ. The function ${}_1\gamma$ contains the information available to a distant observer about what is happening inside the source of the radiation. For this reason it was called by Bondi the *news function*.

We finally mention that in the general case, with no special symmetry assumption, we have two real arbitrary functions combined to a complex ${}_1\gamma$. These are the two degrees of freedom of the gravitational radiation. For the energy of the radiation we have again a formula similar to (49.16), with the quantity $({}_1\gamma_{,0})^2$ replaced by ${}_1\gamma_{,0} \cdot {}_1\bar{\gamma}_{,0}$.

50. Null Tetrads

The use of radiation coordinates has been developed further by Newman and Penrose. The new feature in this development is the systematic use of a *null-tetrad*. In this section we shall give a condensed discussion of the properties of null-tetrads. A brief description of the Newman-Penrose formalism will be given in the next section.

In the 4-dimensional space-time a tetrad is a system of 4 linearly independent vectors z_m^μ; μ is the vector index while $m = 0$, 1, 2 and 3 is an index numbering the 4 vectors. The tetrad is to be given at every point of the space-time (or in a certain region of it) in a continuous way and so that first and second derivatives of the components z_m^μ with respect to the coordinates x^ν exist. The tetrad is characterized by the scalar products ε_{lm} of all possible pairs of vectors:

$$\varepsilon_{lm} = g_{\lambda\mu} z_l^\lambda z_m^\mu = z_{l\mu} z_m^\mu = \varepsilon_{ml} . \tag{50.1}$$

Taking the determinants of both sides of Equation (50.1) we find:

$$\det \varepsilon_{lm} = \det g_{\lambda\mu} \cdot (\det z_m^\mu)^2 . \tag{50.2}$$

Since the vectors z_m^μ are linearly independent, we shall have:

$$\det z_m^\mu \neq 0 .$$

We also have to demand $\det g_{\lambda\mu} \neq 0$ and consequently

$$\det \varepsilon_{lm} \neq 0 .$$

It follows that we can define quantities ε^{lm} by the relation:

$$\varepsilon_{ln}\varepsilon^{mn} = \delta_l^m . \tag{50.3}$$

We can now raise and lower latin indices as well:

$$z^{m\mu} = \varepsilon^{mn} z_n^\mu ; \qquad z_{m\mu} = \varepsilon_{mn} z^n{}_\mu = \varepsilon_{mn} g_{\mu\nu} z^{n\nu} \quad \text{etc.} \tag{50.4}$$

The matrices ε_{mn} and ε^{mn} play a role analogous to $g_{\mu\nu}$ and $g^{\mu\nu}$. Indeed, one can easily prove the formula:

$$g_{\lambda\mu} = \varepsilon_{lm} z^l{}_\lambda z^m{}_\mu , \tag{50.5}$$

which is symmetric to (50.1) with respect to Greek and Latin indices. Equation (50.5) can be written also in the form:

$$\delta_\mu^\lambda = z_m^\lambda z^m{}_\mu \tag{50.6}$$

which is symmetric to (50.1) written in the form:

$$\delta_m^l = z^l_\mu z_m^\mu . \tag{50.6'}$$

A simple tetrad in the space-time is the *orthogonal* one, in which the matrix ε_{lm} is diagonal and has the following normal form:

$$\varepsilon_{lm} = \begin{pmatrix} 1 & 0 & 0 & 0 \\ 0 & -1 & 0 & 0 \\ 0 & 0 & -1 & 0 \\ 0 & 0 & 0 & -1 \end{pmatrix} = \varepsilon^{lm} .$$

It is, however, easy to see that this type of tetrad is not particularly useful in the discussion of the radiation problem. Indeed, in this discussion we have to use radiation coordinates, or equivalently a family of null-surfaces. The null-vector l^μ, which is normal to these surfaces, is evidently of importance and should be taken as one of the vectors of the tetrad:

$$z_0^\mu = l^\mu . \tag{50.7}$$

Since this is a null vector, we shall have:

$$\varepsilon_{00} = 0 \tag{50.8}$$

and consequently the tetrad will not be an orthonormal one.

In what follows we shall consider the more general case in which l^μ is a null vector, but not necessarily normal to a null-surface. The relation (50.8) remains valid. It will be convenient to take as the second vector of the tetrad another null vector:

$$z_1^\mu = n^\mu ; \qquad n_\mu n^\mu = 0 . \tag{50.9}$$

Therefore:

$$\varepsilon_{11} = 0 . \tag{50.10}$$

The remaining two vectors of the tetrad,

$$z_2^\mu = a^\mu, \qquad z_3^\mu = b^\mu,$$

will be space-like and the conditions of orthogonality and of normalization imposed on the 4 vectors are expressed by the following form of the matrix ε_{lm}:

$$\varepsilon_{lm} = \begin{pmatrix} 0 & 1 & 0 & 0 \\ 1 & 0 & 0 & 0 \\ 0 & 0 & -1 & 0 \\ 0 & 0 & 0 & -1 \end{pmatrix} = \varepsilon^{lm}.$$

For the actual calculations it is advantageous to introduce a complex null vector m^μ by the equation:

$$m^\mu = \frac{1}{\sqrt{2}}(a^\mu + ib^\mu). \tag{50.11}$$

One verifies at once the relations:

$$m_\mu m^\mu = 0 = \bar{m}_\mu \bar{m}^\mu; \qquad m_\mu \bar{m}^\mu = -1. \tag{50.12}$$

We choose then the last two vectors of the tetrad rather as follows:

$$z_2^\mu = m^\mu, \qquad z_3^\mu = \bar{m}^\mu. \tag{50.13}$$

We thus have the *null-tetrad* composed of the two real null vectors l^μ and n^μ and the complex-conjugate null vectors m^μ and $\bar{m}^\mu$. The matrix ε_{lm} is now:

$$\varepsilon_{lm} = \begin{pmatrix} 0 & 1 & 0 & 0 \\ 1 & 0 & 0 & 0 \\ 0 & 0 & 0 & -1 \\ 0 & 0 & -1 & 0 \end{pmatrix} = \varepsilon^{lm}. \tag{50.14}$$

With the help of a tetrad we derive from any tensor a set of scalars. For example from the tensor $T^{\lambda\mu}$:

$$T^{\lambda\mu} z_{l\lambda} z_{m\mu} \equiv T_{lm}. \tag{50.15}$$

These scalars determine completely the tensor, since we have the inverse relation:

$$T_{lm} z^{l\lambda} z^{m\mu} = T^{\lambda\mu}. \tag{50.16}$$

They are called the *tetrad components* of the tensor. The relation (50.1) shows that the ε_{lm} are the tetrad components of the metric tensor.

From the derivatives of the components of the tetrad vectors one can construct a set of scalars γ_{lmn} defined by the relation:

$$\gamma_{lmn} = z_{l\mu;\nu} z_m^\mu z_n^\nu. \tag{50.17}$$

The *rotation coefficients* γ_{lmn} are, in a certain sense, analogous to the Christoffel symbols. For a tetrad satisfying the condition:

$$\varepsilon_{lm} = \text{const} \tag{50.18}$$

one can derive easily the relations:

$$\gamma_{lmn} + \gamma_{mln} = 0\,. \tag{50.19}$$

Consequently in the case of an orthonormal or of a null tetrad there are 24 independent coefficients γ_{lmn}. In the case of the null tetrad (50.14) the use of the complex null vector m^μ has the consequence that the γ_{lmn} are partly complex. Newman and Penrose obtained, by an approach based on spinor calculus, a systematic complexification of the γ_{lmn} which then reduce to 12 complex *spin coefficients*.

In the case of the null tetrad, which interests us here, the geometrical meaning of the coefficients γ_{lmn} is rather complicated. Only the coefficients

$$\gamma_{023} = l_{\mu;\nu} m^\mu \bar{m}^\nu \equiv \varrho\,, \qquad \gamma_{022} = l_{\mu;\nu} m^\mu m^\nu \equiv \sigma \tag{50.20}$$

have a simple geometrical meaning, as we shall show now. We consider again the case in which the null vector l^μ is not necessarily orthogonal to a null surface. But we shall assume, in order to arrive at relatively simple results, that the vectorfield l^μ defines a *geodesic* null congruence.

We consider a point P and the corresponding vector l^μ (Figure 32). For a more intuitive reasoning we reintroduce the real space-like vectors a^μ and b^μ appearing in (50.11):

$$a^\mu = \frac{1}{\sqrt{2}}(m^\mu + \bar{m}^\mu)\,, \qquad b^\mu = \frac{1}{i\sqrt{2}}(m^\mu - \bar{m}^\mu)\,. \tag{50.21}$$

On the 2-dimensional surface spanned by a^μ and b^μ we consider the circle C with centre P and radius 1. (The exact reasoning requires a small radius ε.)

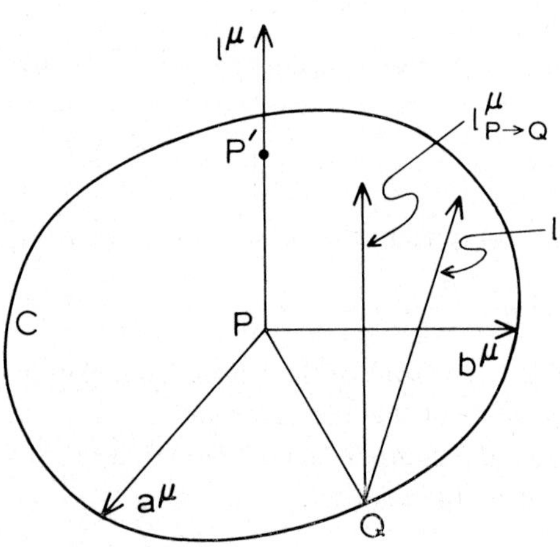

Fig. 32. Determining the geometrical meaning of the coefficients ϱ and σ.

The null geodesics starting from the points of the circle C and tangent to the vector-field l^μ form a cylinder-like tube σ, whose intersection with the 2-dimensional surface determined by a^μ and b^μ at the point P is the circle C. We now ask the following question: What will be the section C' of this tube with the similar 2-dimensional surface corresponding to the neighbouring point P' (Figure 32)?

In order to answer this question we consider at the point Q of the circle C the following two vectors:

(i) The vector l_Q^μ of the field l^μ at Q, and

(ii) the vector $l_{P\to Q}^\mu$ which is the result of the parallel transport of the vector l_P^μ from P to Q along the displacement $\mathbf{PQ} \equiv s^\mu$ (Figure 33). The difference

$$\delta l^\mu = l_Q^\mu - l_{P\to Q}^\mu \tag{50.22}$$

can be determined with the help of Equation (7.5) and is:

$$\delta l^\mu = l^\mu{}_{;\nu} s^\nu . \tag{50.23}$$

The vector δl^μ can be decomposed with the help of the tetrad as follows:

$$\delta l^\mu = a' a^\mu + b' b^\mu + c l^\mu + d n^\mu . \tag{50.24}$$

Since l^μ is a null vector, we have:

$$l_\mu l^\mu{}_{;\nu} = 0 .$$

We therefore find from (50.24) multiplied by l_μ, taking into account (50.14):

$$d = 0 .$$

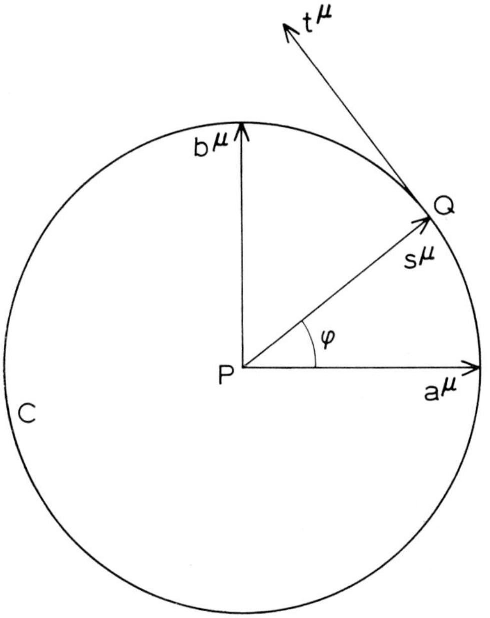

Fig. 33. The angle φ and the vectors s^μ and t^μ.

It will be convenient to use instead of a^μ and b^μ the vectors s^μ and t^μ, t^μ being the unit vector tangent to the circle C at Q (Figure 33). We shall then have:

$$\delta l^\mu = a s^\mu + b t^\mu + c l^\mu . \tag{50.25}$$

One sees at once from Figure 32 that, if a and b vanish, the vectors l^μ_Q and $l^\mu_{P\to Q}$ will be collinear and the cross-section C' of the tube σ at P' will be the result of the parallel transport from P to P' of the cross-section at P, i.e. of the circle C. The change of the form of C' relative to that of C will therefore depend on the coefficients a and b.

In order to determine a and b we multiply (50.25) respectively by s^μ and t^μ. The result is:

$$a = - l_{\mu;\nu} s^\mu s^\nu , \qquad b = - l_{\mu;\nu} t^\mu s^\nu . \tag{50.26}$$

The vectors s^μ and t^μ are expressed in terms of a^μ and b^μ by the relations:

$$s^\mu = a^\mu \cos\varphi + b^\mu \sin\varphi , \qquad t^\mu = - a^\mu \sin\varphi + b^\mu \cos\varphi , \tag{50.27}$$

φ being the angle formed by s^μ and a^μ (Figure 33). Introducing again the complex null vector m^μ we find:

$$s^\mu = \frac{1}{\sqrt{2}} (e^{-i\varphi} m^\mu + e^{i\varphi} \bar{m}^\mu) , \qquad t^\mu = \frac{1}{i\sqrt{2}} (e^{-i\varphi} m^\mu - e^{i\varphi} \bar{m}^\mu) . \tag{50.28}$$

With the values (50.28) of s^μ and t^μ we find from (50.26), taking into account the definitions (50.20) of ϱ and σ:

$$\left.\begin{aligned} a &= - \tfrac{1}{2} (e^{-2i\varphi} \sigma + e^{2i\varphi} \bar{\sigma} + \varrho + \bar{\varrho}) , \\ b &= \frac{i}{2} (e^{-2i\varphi} \sigma - e^{2i\varphi} \bar{\sigma} + \varrho - \bar{\varrho}) . \end{aligned}\right\} \tag{50.29}$$

If we write:

$$\varrho = |\varrho| e^{i\alpha} , \qquad \sigma = |\sigma| e^{i\beta} \tag{50.30}$$

we find finally:

$$\left.\begin{aligned} a &= - |\varrho| \cos\alpha - |\sigma| \cos(2\varphi - \beta) , \\ b &= - |\varrho| \sin\alpha + |\sigma| \sin(2\varphi - \beta) . \end{aligned}\right\} \tag{50.31}$$

Introducing these expressions in (50.25) we find:

$$\begin{aligned} \delta l^\mu - c l^\mu = &- |\varrho| \cos\alpha \cdot s^\mu - |\varrho| \sin\alpha \cdot t^\mu + |\sigma| \{ - \cos(2\varphi - \beta) s^\mu + \\ &+ \sin(2\varphi - \beta) t^\mu \} . \end{aligned} \tag{50.32}$$

The first term on the right-hand side of Equation (50.32) is seen from Figure 33 to describe an *expansion* of the circle C along the null geodesic of the point P. The rate of expansion is:

$$\theta = - |\varrho| \cos\alpha = - \frac{\varrho + \bar{\varrho}}{2} . \tag{50.33}$$

The second term describes a *twist* of the circle, the rate of which is

$$\omega = |\varrho| \sin\alpha = \frac{\varrho - \bar{\varrho}}{2i}. \tag{50.34}$$

The coefficient of $|\sigma|$ in the last term of (50.32) depends on the angle φ. It is easy to see that because of this dependence the circle C goes over into an ellipse with the two axes proportional to $1 \pm \frac{1}{2}|\sigma|\, d\lambda$, $d\lambda$ being the increase of the parameter of the null geodesic from P to P'. Hence $|\sigma|$ represents the *distortion* or *shear* of the circle C.

If the geodesic congruence defined by the vectorfield l^μ is referred to an affine parameter, the geodesic equation has the form (8.8), which is equivalent to:

$$l_{\mu;\nu} l^\nu = 0.$$

In this case the quantities θ, ω and $|\sigma|$ can be expressed in terms of the field l^λ only. This is obtained by the discussion of the tetrad decomposition of the tensor $l_{\mu;\nu}$:

$$l_{\mu;\nu} = a_{mn} z^m{}_\mu z^n{}_\nu.$$

The final formulae are:

$$\left.\begin{aligned} \theta &= \tfrac{1}{2} l^\mu{}_{;\mu}, \\ \omega &= \left\{\tfrac{1}{2} l_{[\mu;\nu]} l^{\mu;\nu}\right\}^{1/2}, \\ |\sigma| &= \left\{\tfrac{1}{2} l_{(\mu;\nu)} l^{\mu;\nu} - \theta^2\right\}^{1/2}. \end{aligned}\right\} \tag{50.35}$$

51. The Method of Newman and Penrose

This method is characterized by the following two features.

(i) A null tetrad is introduced and then used systematically, all tensors and tensor equations being expressed in terms of their tetrad components. The tetrad is, of course, adapted to radiation coordinates: The first vector of the tetrad is the normal to the null surfaces which form the basis of the radiation coordinates:

$$z_{0\mu} = l_\mu = f_{,\mu}. \tag{51.1}$$

(ii) The field equations are all of the first order. This is obtained in the usual way; one considers an enlarged system of field variables and a correspondingly enlarged system of equations.

There are now the following 3 sets of field variables:

(i) The components of the tetrad vectors z^μ_m. These components are equivalent to the metric components $g_{\lambda\mu}$, as one can see from Equation (50.5) which has, because of (50.14), the following form:

$$g_{\lambda\mu} = l_\lambda n_\mu + l_\mu n_\lambda - (m_\lambda \bar{m}_\mu + m_\mu \bar{m}_\lambda). \tag{51.2}$$

(ii) The rotation coefficients γ_{lmn}.

(iii) The tetrad components R_{lmnp} of the Riemann tensor, derived from the tensor components $R_{\lambda\mu\nu\varrho}$ with the help of an equation of the type (50.15):

$$R_{lmnp} = R_{\lambda\mu\nu\varrho} z_l^\lambda z_m^\mu z_n^\nu z_p^\varrho . \tag{51.3}$$

The enlarged system of field equations consists also of 3 sets. The first set is given by the Equations (50.17) which express the γ_{lmn} in terms of the components of the tetrad vectors and their first derivatives.

The second set of field equations are the equations expressing the tetrad components of the Riemann tensor in terms of the γ_{lmn} and their first derivatives. These equations can be obtained from the definition of the Riemann tensor corresponding to (9.2) with $a_\lambda = z_{l\lambda}$:

$$z_{l\lambda;\mu;\nu} - z_{l\lambda;\nu;\mu} = R^\alpha_{\lambda\mu\nu} z_{l\alpha} . \tag{51.4}$$

We need the tetrad components of this relation, which are obtained if we multiply Equation (51.4) by $z_m^\lambda z_n^\mu z_p^\nu$. Taking into account the definition (50.17) of the rotation coefficients we find finally that R_{lmnp} can be expressed in terms of the γ_{lmn} and their first derivatives:

$$R_{lmnp} = \gamma_{lmn:\,p} - \gamma_{lmp:\,n} + \gamma_{lsp}\gamma_{m\ n}^{\ s} - \gamma_{lsn}\gamma_{m\ p}^{\ s} + \gamma_{lms}(\gamma_{n\ p}^{\ s} - \gamma_{p\ n}^{\ s}) . \tag{51.5}$$

The symbol $(\ldots)_{:\,p}$ denotes the derivative in the direction z_p^ϱ:

$$(\ldots)_{:p} = (\ldots)_{,\varrho} z_p^\varrho . \tag{51.6}$$

For actual calculations it is useful to decompose the Riemann tensor as in Section 14, using the Weyl tensor $C_{\lambda\mu\nu\varrho}$, the traceless Ricci tensor $B_{\lambda\mu}$ and the scalar R. This is especially convenient when we have to use the vacuum field equations (19.4), in which case it is sufficient to put

$$B_{\lambda\mu} = 0 = R . \tag{51.7}$$

We have then:

$$R_{\lambda\mu\nu\varrho} = C_{\lambda\mu\nu\varrho} \Leftrightarrow R_{lmnp} = C_{lmnp} . \tag{51.8}$$

There are 10 independent C_{lmnp}, as many as the independent $C_{\lambda\mu\nu\varrho}$. In the formalism which is derived with the help of spinor calculus Newman and Penrose obtain a description of the C_{lmnp} in terms of 5 complex quantities.

The Equations (50.17) and (51.5) are not sufficient in the present situation. We have to consider a third set of equations, which are the tetrad transcription of the classical Bianchi identity. This equation contains first derivatives of the tetrad components R_{lmnp} and terms which are bilinear in R_{lmnp} and γ_{lmn}. It can be obtained by a simple calculation and is:

$$R_{lm[np:\,q]} = \gamma_{l\ [q}^{\ s} R_{np]sm} + \gamma_{m\ [q}^{\ s} R_{np]ls} + 2R_{lms[p}\gamma_{n\ q]}^{\ s} . \tag{51.9}$$

The reason why the relation (51.9) has to be considered now as a field equation is the following. In the present approach we cannot consider the Equations (50.17) and

(51.5) as the equations of definition of γ_{lmn} and R_{lmnp}, because this would lead to equations of second order. We have now to treat all 3 sets of field variables as independent unknowns. Consequently we have to use Equation (51.9), which represents the integrability condition of Equation (51.5), as an independent set of field equations.

Because of the special choice (51.1) of the vector z_0^μ, there are some simplifications of the general results obtained in Section 50. The last part of Equation (51.1) has the following consequence:

$$l_{\mu;\nu} = l_{\nu;\mu}. \tag{51.10}$$

We then find from the first of (50.20):

$$\varrho = \bar{\varrho} \quad \Leftrightarrow \quad \omega = 0. \tag{51.11}$$

Further simplifications are obtained if we choose the tetrad vectors n^μ and m^μ arbitrarily on a 2-dimensional surface S_2 cutting all bicharacteristics of the null-surface Σ and we then determine n^μ and m^μ on the whole Σ by parallel transport along the bicharacteristics.

Another important simplification is obtained from the requirement that all null-surfaces Σ determined by Equation (49.3) tend asymptotically to future light cones of the Minkowski metric to which the metric $g_{\lambda\mu}$ tends for $r \to \infty$. One can determine easily the quantities ϱ and σ for a light cone of the Minkowski space. The result is:

$$\varrho = -\frac{1}{r}, \qquad \sigma = 0.$$

It follows that in the present case we have to demand:

$$\varrho = -\frac{1}{r} + O\left(\frac{1}{r^2}\right). \tag{51.12}$$

A more detailed discussion shows that we can obtain for σ the asymptotic form:

$$\sigma = \frac{\sigma_0}{r^2} + O\left(\frac{1}{r^3}\right), \tag{51.13}$$

where the coefficient σ_0 is independent of r.

The propagation of gravitational radiation far from its sources is discussed again, as in the Bondi method, by assuming that all field quantities can be developed in power series of $1/r$. We shall not enter into the details of this discussion, of which we give only one of the main results: The coefficient σ_0 of the first term of the development (51.13) for σ is the only quantity which remains entirely unrestricted by the field equations. It is equivalent to the news function of Bondi and determines the energy of the gravitational radiation by a formula of the type (49.16):

$$\frac{\mathrm{d}E}{\mathrm{d}t} = \frac{1}{4\pi}\int \dot{\sigma}_0 \dot{\bar{\sigma}}_0 \,\mathrm{d}\omega, \qquad (\,.\,.\,)^{\cdot} = \frac{\partial}{\partial x^0}(\,.\,.\,). \tag{51.14}$$

Exercises

XI1: Verify by computation the relations (45.34).

XI2: Determine the coordinate transformations which eliminate discontinuities $\gamma_{\mu\nu}$ of the form (47.16).

XI3: Show that the trajectories of the null vector p_λ defined, according to the second Equation (47.15), on the null surface Σ are null geodesics of the space-time.

XI4: Discuss gravitational shock waves of order $n=1$. (Demand the vanishing of the surface distribution on Σ of the matter tensor $T_{\lambda\mu}$.)

XI5: Discuss gravitational and electromagnetic shock waves of order $n=2$ in the Einstein-Maxwell theory.

XI6: Prove that the Schwarzschild metric is of the type D. (Use the form (49.13) of the metric.)

XI7: Derive formulae (50.35). (Decompose $l_{\mu;\nu}$ in tetrad components:

$$l_{\mu;\nu} = \alpha^{mn} z_{m\mu} z_{n\nu}$$

and use the restrictions imposed on l_μ:

$$l_\mu l^\mu = 0, \qquad l_{\mu;\nu} l^\nu = 0.)$$

XI8: Determine ϱ and σ for a null plane ($x - t = \text{const}$), a null cylinder ($\sqrt{x^2 + y^2} - t = \text{const}$) and a null cone in Minkowski space.

CHAPTER XII

THE COSMOLOGICAL PROBLEM

52. Historical Survey

The main problem in cosmology is the study of the dynamical structure of the Universe as a whole.

From the physicist's point of view the study of the whole Universe constitutes a problem of an asymptotic character. Indeed, the history of science shows that scientific progress consists in a gradual broadening of the limits of the already explored region of the Universe. The broadening is to be understood in a general sense, as it is proceeding in two opposite directions: Starting from the laboratory dimensions we try to understand the larger scale phenomena in astrophysics, but also the smaller scale in microphysics. On the other side, for the purpose of research the most useful working hypothesis consists in assuming that the Universe is infinite, in extension as well as in variety. Accordingly, arriving at the detailed knowledge of the structure of the whole Universe, which would constitute the complete solution of the cosmological problem, may in principle be attainable, but only asymptotically at $t \to \infty$.

However, the hope to arrive at a qualitative understanding of the large scale features of the Universe exists since the beginning of the systematic development of astronomy and some well defined cosmological questions were formulated already in pre-relativistic physics. It was not possible to give satisfactory answers to these questions because of the difficulties arising from the fact that prerelativistic physics is based on the use of the Euclidean 3-dimensional space. In particular the infinite extension of Euclidean space makes impossible the discussion of the dynamical structure of the Universe on the basis of classical Newtonian mechanics. The first real possibility of discussing the dynamical problem of the Universe was given by general relativity.

Actually there is now also a 'Newtonian cosmology', based on a modification of the concept of the newtonian inertial coordinate systems. It must, however, be remembered that this modification has been made possible only after the successful treatment of the problem according to general relativity.

The history of the first discussion of the cosmological problem in general relativity is interesting and instructive. We already mentioned in Section 19 that Mach's principle played an important role in the formulation of general relativity. The essential content of this principle is that the inertia of a body is the consequence of some global interaction of the body with the material distribution of the whole Universe.

As explained in Section 18, inertial properties are described by the metric. According to the Einstein field equations, any given distribution of matter has an influence on the metric. In the usual *physical* problems – in contrast to the cosmological problem – we are dealing with a bounded material distribution which is assumed to be 'isolated', i.e. to have negligible interaction with the rest of the Universe. In this case the metric is not determined completely by the field equations and the material distribution, since we need also the *boundary conditions* (20.11):

$$g_{\mu\nu} \to \eta_{\mu\nu} \quad \text{for} \quad r \to \infty . \tag{52.1}$$

Consequently, Mach's principle cannot be incorporated completely in the Einstein field equations, when these equations are applied to a physical problem.

It follows from the preceding remarks that the possibility of fully incorporating Mach's principle in the Einstein equations should be expected only in the application of these equations to the cosmological problem. This is really the case, at least for the type of cosmological solutions in which the 3-dimensional space is closed. Indeed, for the solutions of this type the question of boundary conditions does not exist. This qualitative success was one of the reasons for the systematic application of general relativity to the cosmological problem.

The first cosmological model discussed by Einstein was the time-independent or *static* one, which was considered also in prerelativistic physics. The reason for the preference given to this model was that at that time – about 1917 – only small relative velocities of astronomical objects were known.

As we shall see in some detail in Section 53, the quantitative discussion of a cosmological model is based on the statistically averaged distribution of matter, which is assumed to depend on time only. In the static model the average matter tensor has to be constant. Starting from the field equations (19.7),

$$R_{\mu\nu} - \tfrac{1}{2} g_{\mu\nu} R = - \kappa T_{\mu\nu} , \tag{52.2}$$

Einstein found that there was no satisfactory solution corresponding to the static model. The physical reason for this result is the fact that the attractive gravitational forces cannot be balanced by stresses in a way which is compatible with the field equations (52.2).

After this negative result Einstein considered the field equations (19.14) with the cosmological term:

$$R_{\mu\nu} - \tfrac{1}{2} g_{\mu\nu} R + \Lambda g_{\mu\nu} = - \kappa T_{\mu\nu} . \tag{52.3}$$

The result was that there is now a time-independent solution describing a static model. This is because, as mentioned in Section 19, the cosmological term in Equation (52.3) represents an interaction expressed by a repulsive force: We now have attractive as well as repulsive forces and an equilibrium is possible.

However, it was found later that this solution is unstable. The explanation of this instability is straightforward: With increasing distance r of two particles the attractive force decreases as r^{-2}, while the repulsive force increases as r^2. Furthermore, a logical difficulty presented itself when de Sitter found a solution of the Equations (52.3) having no material sources: Einstein had pointed out that in order to have Mach's principle completely incorporated in the field equations, there should be no solution of these equations without sources. We may add that the de Sitter solution had a novel feature which became important later: It describes an *expanding* Universe, in which the distance between two test particles at rest (in the 3-dimensional space) increases with time.

The next two important events were the following:

(i) Friedmann obtained in 1922 solutions of the field equations, with or without cosmological term. These solutions are non-static and represent an expanding or contracting Universe.

(ii) In 1929 the *Hubble effect* was discovered: Distant galaxies show a red-shift of spectral lines which increases nearly proportionally to their distance from the observer. No other explanation of these large shifts has since been found and we have to interpret them as Doppler shifts, showing that actually *the Universe is expanding.*

These two events were decisive in making cosmology a well defined part of theoretical astrophysics.

53. The Cosmological Principle. Form of the Metric

It is not possible to retain in the discussion of the dynamical structure of the Universe as a whole the finer details of the material distribution. Usually the cosmological models are based on the so called *cosmological principle*, according to which the material distribution in the Universe is *homogeneous* and *isotropic*. These assumptions do correspond to the actual observations, on condition that we form statistically averaged values over sufficiently large regions. Indeed, there are important inhomogeneities of the matter distribution, when we consider it on a relatively small scale: Stars form galaxies, galaxies form clusters of galaxies etc. However, on a large scale we find a statistically homogeneous distribution of galaxies and no preferred directions, even out to the maximum distances which can be reached with the instruments available at present.

Taking into account these results we assume that the average matter tensor is independent of the spatial coordinates x^i. It can depend only on the time t, when we are considering a time-dependent model. Further we conclude from the observed isotropy of the large scale distribution of matter that there is no rotation of the Universe as a whole. It follows that every point of the 3-dimensional space has to be equivalent to any other point. The mathematical expression of this requirement is found to be given by the statement that the 3-dimensional space will be a space of *constant curvature.*

The equation characterizing a space of constant curvature is:

$$R_{\lambda\mu\nu\varrho} = k(g_{\lambda\nu}g_{\mu\varrho} - g_{\lambda\varrho}g_{\mu\nu}), \qquad k = \text{const}. \tag{53.1}$$

We have to apply Equation (53.1) to the 3-dimensional subspace of the space-time. In order to avoid confusion with similar quantities of the space-time we shall denote the Riemann tensor of the 3-dimensional space by r_{iklm} and the metric by γ_{ik}. Thus we shall have:

$$r_{iklm} = k(\gamma_{il}\gamma_{km} - \gamma_{im}\gamma_{kl}). \tag{53.2}$$

In the case of a 3-dimensional space the Ricci tensor is equivalent to the Riemann tensor, since each one of these tensors has 6 independent components. Contracting (53.2) with γ^{im} we find:

$$r_{kl} = -2k\gamma_{kl}. \tag{53.3}$$

Equation (53.3) is equivalent to (53.2) and consequently we can use it for determining the form of γ_{kl}.

We first remark that the 3-dimensional space of constant curvature is spherically symmetric and each of its points may be taken as the origin of the coordinate system. It follows that the line element of this space will have the form of the 3-dimensional part of (20.15):

$$\left.\begin{aligned} d\sigma^2 &= \gamma_{ik}\, dx^i\, dx^k = e^{\mu}\, dr^2 + r^2(d\theta^2 + \sin^2\theta\, d\varphi^2), \\ \mu &= \mu(r). \end{aligned}\right\} \tag{53.4}$$

We now determine the Ricci tensor of the space with the metric (53.4). By a straightforward calculation, which is similar to that given in Section 21, we find:

$$\left.\begin{aligned} &r_{11} = -\frac{1}{r}\mu', \qquad r_{22} = \frac{r_{33}}{\sin^2\theta} = e^{-\mu} - 1 - \frac{r}{2}e^{-\mu}\mu'; \\ &r_{ik} = 0 \quad \text{for} \quad i \neq k. \end{aligned}\right\} \tag{53.5}$$

The relation (53.3) gives the following two equations:

$$-\frac{1}{r}\mu' = -2ke^{\mu}, \qquad e^{-\mu} - 1 - \frac{r}{2}e^{-\mu}\mu' = -2kr^2. \tag{53.6}$$

The solution of these equations is found at once to be:

$$e^{-\mu} = 1 - kr^2. \tag{53.7}$$

Consequently the metric of the 3-dimensional space of constant curvature is:

$$d\sigma^2 = \frac{dr^2}{1 - kr^2} + r^2(d\theta^2 + \sin^2\theta\, d\varphi^2). \tag{53.8}$$

The constant k has, from Equation (53.3), the dimension (length)$^{-2}$:

$$k = \varepsilon r_0^{-2}, \tag{53.9}$$

ε taking the values ± 1 or 0 and r_0 being some constant length. We can therefore write Equation (53.8) also in the form:

$$d\sigma^2 = \frac{dr^2}{1 - \varepsilon (r/r_0)^2} + r^2(d\theta^2 + \sin^2\theta\, d\varphi^2). \tag{53.10}$$

For $\varepsilon = +1$ we have a *closed* space of constant curvature. For $\varepsilon = 0$ the space is open and flat. For $\varepsilon = -1$ we have an open space of negative curvature.

We shall assume that the (average) matter tensor $T_{\lambda\mu}$ has the simple form (28.4) corresponding to a perfect fluid:

$$T_{\lambda\mu} = (\varrho + p)\, u_\lambda u_\mu - p g_{\lambda\mu}, \tag{53.11}$$

the quantities ϱ and p depending on t only:

$$\varrho = \varrho(t), \qquad p = p(t). \tag{53.12}$$

We shall use comoving coordinates, which have been defined in Section 28. In these coordinates we have the relations (28.13) and (28.17):

$$u^\mu = (1, 0, 0, 0). \tag{53.13}$$

The metric will be of the form (28.16), simplified by the fact that the 3-dimensional space is of constant curvature. The result is:

$$\mathrm{d}s^2 = \mathrm{d}t^2 - \{F(t)\}^2\, \mathrm{d}\sigma^2, \tag{53.14}$$

the 3-dimensional line element $\mathrm{d}\sigma^2$ being given by (53.10). Note that in (53.14) the factor $F(t)$ has no physical dimension. It will be more convenient to use the dimensionless radial variable ζ defined by the relation:

$$\zeta = \frac{r}{r_0}. \tag{53.15}$$

We shall then have:

$$\mathrm{d}s^2 = dt^2 - R^2\, \mathrm{d}\tilde{\sigma}^2\,; \qquad R = R(t) = r_0 F(t), \tag{53.16}$$

$$\mathrm{d}\tilde{\sigma}^2 = \frac{\mathrm{d}\zeta^2}{1 - \varepsilon\zeta^2} + \zeta^2 (\mathrm{d}\theta^2 + \sin^2\theta\, \mathrm{d}\varphi^2). \tag{53.17}$$

54. Some Cosmological Solutions

We shall calculate the Ricci tensor of the 4-dimensional metric (53.16) in order to write down the field equations. We give only some intermediate results. For the Christoffel symbols $\Gamma^\lambda_{\mu\nu}$ of the space-time we find:

$$\left.\begin{aligned} &\Gamma^i_{kl} = \tilde{\gamma}^i_{kl}, \qquad \Gamma^0_{kl} = R\dot{R}\tilde{\gamma}_{kl}, \qquad \Gamma^k_{0l} = \frac{\dot{R}}{R}\,\delta^k_l; \\ &\Gamma^k_{00} = \Gamma^0_{0k} = \Gamma^0_{00} = 0. \end{aligned}\right\} \tag{54.1}$$

$\tilde{\gamma}^i_{kl}$ are the Christoffel symbols of the 3-dimensional metric $\tilde{\gamma}_{kl}$ appearing in the relation (53.17). Further we find:

$$R_{00} = \frac{3}{R}\ddot{R}, \qquad R_{ok} = 0, \qquad R_{kl} = \tilde{r}_{kl} - \tilde{\gamma}_{kl}(R\ddot{R} + 2\dot{R}^2). \tag{54.2}$$

Comparing (53.8) and (53.17) we see that the expression for $d\tilde{\sigma}^2$ is deduced from that for $d\sigma^2$ by the substitution:

$$r \to \zeta, \qquad k \to \varepsilon.$$

Consequently we have from (53.3):

$$\tilde{r}_{kl} = -2\varepsilon\tilde{\gamma}_{kl}. \tag{54.3}$$

The last of Equations (54.2) takes then the form:

$$R_{kl} = -\tilde{\gamma}_{kl}(R\ddot{R} + 2\dot{R}^2 + 2\varepsilon). \tag{54.4}$$

In order to write the field equations (52.3) we need the scalar $R = g^{\mu\nu}R_{\mu\nu}$. With the form (53.16) of the metric we find:

$$g^{\mu\nu}R_{\mu\nu} = R_{00} - \frac{1}{R^2}\tilde{\gamma}^{kl}R_{kl}. \tag{54.5}$$

Since

$$-\tilde{\gamma}^{kl}R_{kl} = 3(R\ddot{R} + 2\dot{R}^2 + 2\varepsilon),$$

we have finally

$$g^{\mu\nu}R_{\mu\nu} = \frac{6}{R^2}(R\ddot{R} + \dot{R}^2 + \varepsilon). \tag{54.6}$$

Introducing the expressions (54.2) and (54.6) into the field equations (52.3) and taking into account the form of the matter tensor determined by (53.11) and (53.13) we find the following two equations:

$$\left.\begin{aligned} &\frac{3}{R^2}(\dot{R}^2 + \varepsilon) - \Lambda = \kappa\varrho, \\ &\frac{1}{R^2}(2R\ddot{R} + \dot{R}^2 + \varepsilon) - \Lambda = -\kappa p. \end{aligned}\right\} \tag{54.7}$$

We shall discuss the Equations (54.7) for some special cases.

The first cosmological solution of Einstein corresponds to

$$\Lambda = 0 \quad \text{and} \quad \dot{R} = 0.$$

From (54.7) we find:

$$\kappa\varrho = \frac{3\varepsilon}{R^2}, \qquad \kappa p = \frac{-\varepsilon}{R^2}. \tag{54.8}$$

We must demand $\varrho > 0$ and therefore

$$\varepsilon = 1.$$

But then we have $p < 0$, i.e. a negative pressure, which is physically unacceptable.

In the case of the Einstein static solutions with $\Lambda \neq 0$ we have again $\dot{R}=0$ and consequently from (54.7):

$$\kappa\varrho = \frac{3\varepsilon}{R^2} - \Lambda, \qquad \kappa p = \frac{-\varepsilon}{R^2} + \Lambda. \tag{54.9}$$

If we assume that $p \ll \varrho$ and neglect p we find:

$$\Lambda \approx \frac{\varepsilon}{R^2}, \qquad \kappa\varrho \approx \frac{2\varepsilon}{R^2} \approx 2\Lambda.$$

Consequently we must have $\varepsilon=1$ and $\Lambda>0$. The cosmological constant is in this model proportional to the (average) density ϱ:

$$\Lambda \approx \frac{\kappa}{2}\varrho.$$

The de Sitter solution is obtained if we put

$$\varrho = p = 0.$$

The field equations (54.7) reduce then to:

$$3\dot{R}^2 + 3\varepsilon - \Lambda R^2 = 0 = 2R\ddot{R} + \dot{R}^2 + \varepsilon - \Lambda R^2. \tag{54.10}$$

In the case $\varepsilon=0$ (flat 3-dimensional space) we find the simple solution:

$$R = \text{const}\cdot e^{t(\Lambda/3)^{1/2}}. \tag{54.11}$$

We now consider the Friedmann solution with $\Lambda=0$. The field equations (54.7) reduce to:

$$\frac{\dot{R}^2 + \varepsilon}{R^2} = \frac{\kappa\varrho}{3}, \qquad \frac{2R\ddot{R} + \dot{R}^2 + \varepsilon}{R^2} = -\kappa p. \tag{54.12}$$

These equations can be integrated easily if we assume that the pressure is negligible. Putting $p=0$ we find from the second Equation (54.12):

$$2R\ddot{R} + \dot{R}^2 + \varepsilon = 0.$$

The first integral of this equation is:

$$R(\dot{R}^2 + \varepsilon) = \text{const} \equiv C. \tag{54.13}$$

We then derive from the first Equation (54.12):

$$\kappa\varrho R^3 = 3C. \tag{54.14}$$

Equation (54.13) can be written in the form:

$$\dot{R}^2 = \frac{C}{R} - \varepsilon. \tag{54.15}$$

The integration of this equation determines completely the cosmological model. We notice that for $\varepsilon = +1$, corresponding to a closed 3-dimensional space, there is a maximum value of R:

$$R_{\max} = C. \tag{54.16}$$

We therefore have an initial expansion turning, at $R = R_{\max}$, to a contraction. There is a singularity at $R = 0$, in the past as well as in the future. For both the cases $\varepsilon = 0$ or -1 (open 3-dimensional spaces) we have either an expansion only with a singularity in the past, or a contraction only with a singularity in the future.

Whether these singularities have a physical meaning or whether they indicate that we have to change the field equations at very high concentrations of matter, is an open question.

We shall examine briefly the red shifts which are to be expected in an expanding Universe. The exact discussion of this question turns out to be a very difficult problem. One considers usually the case of a small distance D between the source of light and the observer, this being by definition a distance for which the relative redshift $\delta\lambda/\lambda$ is small:

$$\frac{\delta\lambda}{\lambda} \ll 1.$$

To the first order in D one arrives at the following formula:

$$\frac{\delta\lambda}{\lambda} = \frac{\dot{R}(t_0)}{R(t_0)} \frac{D}{c} + \cdots. \tag{54.17}$$

$R(t)$ is the function appearing in the metric (53.16) and t_0 the moment at which the spectral line is observed.

The *Hubble constant* H is defined by the relation:

$$\frac{\delta\lambda}{\lambda} = H \frac{D}{c}. \tag{54.18}$$

Comparing (54.17) and (54.18) we find:

$$\frac{\dot{R}(t_0)}{R(t_0)} = H. \tag{54.19}$$

The accepted value of H is at present:

$$H \approx 2 \times 10^{-18}\ \mathrm{s}^{-1}. \tag{54.20}$$

We end this chapter with the following remark. If we knew the (average) density of matter $\varrho = \varrho(t_0)$ with sufficient accuracy, we could determine from the first Equation (54.12) the value of ε. Indeed, we find from this equation, if we reintroduce the omitted factors c:

$$\frac{\varepsilon}{R^2} = \frac{\kappa\varrho c^2}{3} - \frac{\dot{R}^2}{c^2 R^2}.$$

With the values (19.13) for κ and (54.19) for $\dot{R}/R$ we have:

$$\frac{\varepsilon}{R^2} = \frac{1}{c^2}\left(\frac{8\pi}{3} G\varrho - H^2\right). \tag{54.21}$$

The right-hand side of Equation (54.21) vanishes for the value ϱ_{cr} of ϱ which satisfies the equation:

$$\frac{8\pi}{3} G\varrho_{cr} = H^2. \tag{54.22}$$

With the value (54.20) of H we find:

$$\varrho_{cr} \approx 10^{-29} \text{ g cm}^{-3}.$$

According to (54.21) we shall have a closed space, $\varepsilon = 1$, if $\varrho > \varrho_{cr}$. On the contrary, the space will be open, $\varepsilon = 0$ or 1, if $\varrho \leqslant \varrho_{cr}$.

One accepts at present the following lower limit of the density ϱ:

$$\varrho_{min} \approx 2 \times 10^{-31} \text{ g cm}^{-3},$$

which is not very small compared with ϱ_{cr}. An upper limit ϱ_{max} can be obtained only with the help of some more or less uncertain assumptions. It follows that the information available at present does not allow to determine the value of ε.

Exercises

XIII1: In the Euclidean space with $n = 4$,

$$ds^2 = (d\xi^1)^2 + (d\xi^2)^2 + (d\xi^3)^2 + (d\xi^4)^2,$$

consider the hypersphere S defined by the equation:

$$(\xi^1)^2 + (\xi^2)^2 + (\xi^3)^2 + (\xi^4)^2 = R^2, \qquad R = \text{const}.$$

Show that the metric on S has the form (53.10) with $\varepsilon = 1$.

XII2: Determine the transformation $r = f(r')$ bringing the metric (53.10) into the isotropic form:

$$d\sigma^2 = \varphi(r')\{dr'^2 + r'^2(d\theta^2 + \sin^2\theta\, d\varphi^2)\},$$

determining also the function $\varphi(r')$.

INDEX OF SUBJECTS